KEY CONCEPTS & TECHNIQUES IN GIS

JOCHEN ALBRECHT

KEY CONCEPTS & TECHNIQUES IN GIS

SAGE Publications
Los Angeles · London · New Delhi · Singapore

© Jochen Albrecht 2007

First published 2007

 SAGE Publications Ltd
1 Oliver's Yard
55 City Road
London EC1Y 1SP

SAGE Publications Inc.
2455 Teller Road
Thousand Oaks, California 91320

SAGE Publications India Pvt Ltd
B1/I I Mohan Cooperative Industrial Area
Mathura Road, New Delhi 110 044
India

SAGE Publications Asia-Pacific Pte Ltd
33 Pekin Street #02-01
Far East Square
Singapore 048763

Library of Congress Control Number 2007922921

British Library Cataloguing in Publication data

A catalogue record for this book is available from
the British Library

ISBN 978-1-4129-1015-6
ISBN 978-1-4129-1016-3 (pbk)

Typeset by C&M Digitals (P) Ltd, Chennai, India
Printed and bound in Great Britain by TJ International Ltd
Printed on paper from sustainable resources

Contents

List of Figures

Preface

GIS has been coming of age. Millions of people use one GIS or another every day, and with the advent of **Web 2.0** we are promised GIS functionality on virtually every desktop and web-enabled cellphone. GIS knowledge, once restricted to a few insiders working with minicomputers that, as a category, don't exist any more, has proliferated and is bestowed on students at just about every university and increasingly in community colleges and secondary schools. GIS textbooks abound and in the course of 20 years have moved from specialized topics (Burrough 1986) to general-purpose textbooks (Maantay and Ziegler 2006). With such a well-informed user audience, who needs yet another book on GIS?

The answer is two-fold. First, while there are probably millions who use GIS, there are far fewer who have had a systematic introduction to the topic. Many are self-trained and good at the very small aspect of GIS they are doing on an everyday basis, but they lack the bigger picture. Others have learned GIS somewhat systematically in school but were trained with a particular piece of software in mind – and in any case were not made aware of modern methods and techniques. This book also addresses decision-makers of all kinds – those who need to decide whether they should invest in GIS or wait for GIS functionality in Google Earth (Virtual Earth if you belong to the other camp).

This book is indebted to two role models. In the 1980s, Sage published a tremendously useful series of little green paperbacks that reviewed quantitative methods, mostly for the social sciences. They were concise, cheap (as in extremely good quality/price ratio), and served students and practitioners alike. If this little volume that you are now holding contributes to the revival of this series, then I consider my task to be fulfilled. The other role model is an unsung hero, mostly because it served such a small readership. The CATMOG (*Concepts and Techniques in Modern Geography*) series fulfills the same set of criteria and I guess it is no coincidence that it too has been published by Sage. CATMOG is now unfortunately out of print but deserves to be promoted to the modern GIS audience at large, which as I pointed out earlier, is just about everybody. With these two exemplars of the publishing pantheon in house, is it a wonder that I felt honored to be invited to write this volume? My kudos goes to the unknown editors of these two series.

Jochen Albrecht

1   Creating Digital Data

The creation of spatial data is a surprisingly underdeveloped topic in **GIS** literature. Part of the problem is that it is a lot easier to talk about tangibles such as data as a commodity, and **digitizing** procedures, than to generalize what ought to be the very first step: an analysis of what is needed to solve a particular geographic question. Social sciences have developed an impressive array of methods under the umbrella of research design, originally following the lead of experimental design in the natural sciences but now an independent body of work that gains considerably more attention than its counterpart in the natural sciences (Mitchell and Jolley 2001).

For **GIScience**, however, there is a dearth of literature on the proper development of (applied) research questions; and even outside academia there is no vendor-independent guidance for the GIS entrepreneur on setting up the databases that off-the-shelf software should be applied to. GIS vendors try their best to provide their customers with a starter package of basic data; but while this suffices for training or tutorial purposes, it cannot substitute for in-house data that is tailored to the needs of a particular application area.

On the academic side, some of the more thorough introductions to GIS (e.g. Chrisman 2002) discuss the history of spatial thought and how it can be expressed as a dialectic relationship between absolute and relative notions of space and time, which in turn are mirrored in the two most common spatial representations of **raster** and **vector GIS**. This is a good start in that it forces the developer of a new GIS database to think through the limitations of the different ways of storing (and acquiring) spatial data, but it still provides little guidance.

One of the reasons for the lack of literature – and I dare say academic research – is that far fewer GIS would be sold if every potential buyer knew how much work is involved in actually getting started with one's own data. Looking from the ivory tower, there are ever fewer theses written that involve the collection of relevant data because most good advisors warn their mentees about the time involved in that task and there is virtually no funding of basic research for the development of new methods that make use of new technologies (with the exception of **remote sensing** where this kind of research is usually funded by the manufacturer). The GIS trade magazines of the 1980s and early 90s were full of eye-witness reports of GIS projects running over budget; and a common claim back then was that the development of the database, which allows a company or regional authority to reap the benefits of the investment, makes up approximately 90% of the project costs. Anecdotal evidence shows no change in this staggering character of GIS data assembly (Hamil 2001).

So what are the questions that a prospective GIS manager should look into before embarking on a GIS implementation? There is no definitive list, but the following questions will guide us through the remainder of this chapter.

- What is the nature of the data that we want to work with?
- Is it quantitative or qualitative?
- Does it exist hidden in already compiled company data?
- Does anybody else have the data we need? If yes, how can we get hold of it? See also Chapter 2.
- What is the scale of the phenomenon that we try to capture with our data?
- What is the size of our study area?
- What is the resolution of our sampling?
- Do we need to update our data? If yes, how often?
- How much data do we need, i.e. a sample or a complete census?
- What does it cost? An honest cost–benefit analysis can be a real eye-opener.

Although by far the most studied, the first question is also the most difficult one (Gregory 2003). It touches upon issues of research design and starts with a set of goals and objectives for setting up the GIS database. What are the questions that we would like to get answered with our GIS? How immutable are those questions – in other words, how flexible does the setup have to be? It is a lot easier (and hence cheaper) to develop a database to answer one specific question than to develop a general-purpose system. On the other hand, it usually is very costly and sometimes even impossible to change an existing system to answer a new set of questions.

The next step is then to determine what, in an ideal world, the data would look like that answers our question(s). Our world is not ideal and it is unlikely that we will gather the kind of data prescribed in this step, but it is interesting to understand the difference between what we would like to have and what we actually get. Chapter 3 will expand on the issues related to imperfect data.

1.1 Spatial data

In its most general form, geographic data can be described as any kind of data that has a **spatial reference**. A spatial reference is a descriptor for some kind of location, either in direct form expressed as a **coordinate** or an **address** or in indirect form relative to some other location. The location can (1) stand for itself or (2) be part of a spatial object, in which case it is part of the boundary definition of that object.

In the first instance, we speak of a **field view** of geographic information because all the **attribute**s associated with that location are taken to accurately describe everything at that very position but are to be taken less seriously the further we get away from that location (and the closer we can to another location).

The second type of locational reference is used for the description of **geographic objects**. The position is part of a geometry that defines the boundary of that object.

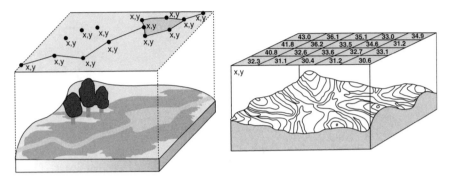

Figure 1   Object vs. field view (vector vs. raster GIS)

The attributes associated with this piece of geographic data are supposed to be valid for all coordinates that are part of the geographic object. For example, if we have the attribute 'population density' for a census unit, then the density value is assumed to be valid throughout this unit. This would obviously be unrealistic in the case where a quarter of this unit is occupied by a lake, but it would take either lots of auxiliary information or sophisticated techniques to deal with this representational flaw. Temporal aspects are treated just as another attribute. GIS have only very limited abilities to reason about temporal relationships.

This very general description of spatial data is slightly idealistic (Couclelis 1992). In practice, most GIS distinguish strictly between the two types of spatial perspectives – the field view that is typically represented using raster GIS, versus the **object view** exemplified by vector GIS (see Figure 1). The sets of functionalities differ considerably depending on which perspective is adopted.

1.2 Sampling

But before we get there, we will have to look at the relationship between the real-world question and the technological means that we have to answer it. Helen Couclelis (1982) described this process of abstracting from the world that we live in to the world of GIS in the form of a 'hierarchical man' (see Figure 2). GIS store their spatial data in a two-dimensional Euclidean geometry representation, and while even spatial novices tend to formalize geographic concepts as simple geometry, we all realize that this is not an adequate representation of the real world. The hierarchical man illustrates the difference between how we perceive and conceptualize the world and how we represent it on our computers. This in turn then determines the kinds of questions (procedures) that we can ask of our data.

This explains why it is so important to know what one wants the GIS to answer. It starts with the seemingly trivial question of what area we should collect the data for – 'seemingly' because, often enough, what we observe for one area is influenced by factors that originate from outside our area of interest. And unless we have

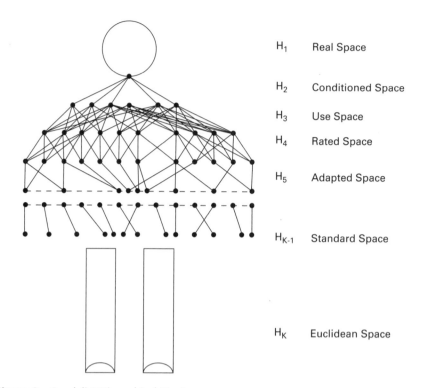

H_1	Real Space
H_2	Conditioned Space
H_3	Use Space
H_4	Rated Space
H_5	Adapted Space
H_{K-1}	Standard Space
H_K	Euclidean Space

Figure 2 Couclelis' 'Hierarchical Man'

complete control over all aspects of all our data, we might have to deal with bound-aries that are imposed on us but have nothing to do with our research question (the modifiable area unit problem, or **MAUP**, which we will revisit in Chapter 10). An example is street crime, where our outer research boundary is unlikely to be related to the city boundary, which might have been the original research question, and where the reported cases are distributed according to police precincts, which in turn would result in different spatial statistics if we collected our data by precinct rather than by address (see Figure 3).

In 99% of all situations, we cannot conduct a complete census – we cannot inter-view every customer, test every fox for rabies, or monitor every brown field (former industrial site). We then have to conduct a sample and the techniques involved are radically different depending on whether we assume a discrete or continuous distri-bution and what we believe the causal factors to be. We deal with a chicken-and-egg dilemma here because the better our understanding of the research question, the more specific and hence appropriate can be our sampling technique. Our needs, however, are exactly the other way around. With a generalist ('if we don't know any-thing, let's assume random distribution') approach, we are likely to miss the crucial events that would tell us more about the unknown phenomenon (be it West Nile virus or terrorist chatter).

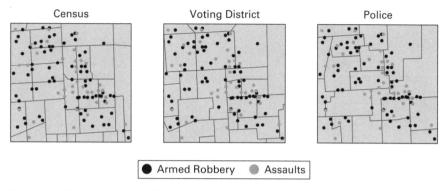

Figure 3 Illustration of variable source problem

Most sampling techniques apply to so-called point data; i.e., individual locations are sampled and assumed to be representative for their immediate neighborhood. Values for non-sampled locations are then interpolated assuming continuous distributions. The interpolation techniques will be discussed in Chapter 10. Currently unresolved are the sampling of discrete phenomena, and how to deal with spatial distributions along networks, be they river or street networks.

Surprisingly little attention has been paid to the appropriate scale for sampling. A neighborhood park may be the world to a squirrel but is only one of many possible hunting grounds for the falcon nesting on a nearby steeple (see Figure 4). Every geographic phenomenon can be studied at a multitude of scales but usually only a small fraction of these is pertinent to the question at hand. As mentioned earlier, knowing what one is after goes a long way in choosing the right approach.

Given the size of the study area, the assumed form of spatial distribution and scale, and the budget available, one eventually arrives at a suitable spatial resolution. However, this might be complicated by the fact that some spatial distributions change over time (e.g. people on the beach during various seasons). In the end, one has to make sure that one's sampling represents, or at least has a chance to represent, the phenomenon that the GIS is supposed to serve.

1.3 Remote sensing

Without wasting too much time on the question whether remotely sensed data is primary or secondary data, a brief synopsis of the use of image analysis techniques as a source for spatial data repositories is in order. Traditionally, the two fields of GIS and remote sensing were cousins who acknowledged each other's existence but otherwise stayed clearly away from each other. The widespread availability of remotely sensed data and especially pressure from a range of application domains have forced the two communities to cross-fertilize. This can be seen in the added functionalities of both GIS and remote sensing packages, although the burden is still on the user to extract information from remotely sensed data.

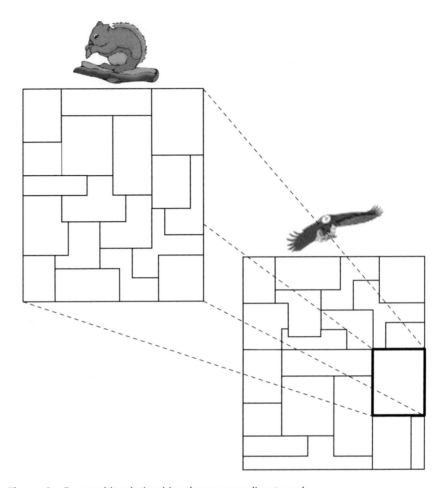

Figure 4 Geographic relationships change according to scale

Originally, GIS and remote sensing data were truly complimentary by adding context to the respective other. GIS data helped image analysts to classify otherwise ambiguous pixels, while imagery used as backdrop to highly specialized vector data provides orientation and situational setting. Truly integrated software that mixes and matches raster, vector and image data for all kinds of GIS functions does not exist; at best, some raster analytical functions take vector data as determinants of processing boundaries. To make full use of remotely sensed data, the GIS user needs to understand the characteristics of a wide range of sensors and what kind of manipulation the imagery has undergone before it arrives on the user's desk.

Remotely sensed data is a good example for the field view of spatial information discussed earlier. For each location we are given a value, called **digital number** (DN), usually in the range from 0 to 255, sometimes up to 65,345. These digital numbers are visualized by different colors on the screen but the software works with DN values rather than with colors. The satellite or airborne sensors have different sensitivities in a wide range of the electromagnetic spectrum, and one aspect that is confusing for many GIS users is that the relationship between a color on the screen and a DN representing a particular but very small range of the electromagnetic spectrum is arbitrary. This is unproblematic as long as we leave the analysis entirely to the computer – but there is only a very limited range of tasks that can be performed automatically. In all other instances we need to understand what a screen color stands for.

Most remotely sensed data comes from so-called passive sensors, where the sensor captures reflections of energy of the earth's surface that originally comes from the sun. Active sensors on the other hand send their own signal and allow the image analyst to make sense of the difference between what was sent off and what bounces back from the 'surface'. In either instance, the word *surface* refers either to the topographic surface or to parts in close vicinity, such as leaves, roofs, minerals or water in the ground. Early generations of sensors captured reflections predominantly in a small number of bands of the visible (to the human eye) and infrared ranges, but the number of spectral bands as well as their distance from the visible range has increased. In addition, the resolution of images has improved from multiple kilometers to fractions of a meter (or centimeters in the case of airborne sensors).

With the right sensor, software and expertise of the operator we can now use remotely sensed data to distinguish not only various kinds of crops but also their maturity, response to drought conditions or mineral deficiencies. We can detect buried archaeological sites, do mineral exploration, and measure the height of waves. But all of these require a thorough understanding of what each sensor can and cannot capture as well as what conceptual model image analysts use to draw their conclusions from the digital numbers mentioned above. The difference between academic theory and operational practice is often discouraging. This author, for instance, searched in vain for imagery that helps to discern the vanishing rate of Irish bogs because for many years there happened to be no coincidence between cloudless days and a satellite over these areas on a clear day.

On the upside, once one has the kind of remotely sensed data that the GIS practitioner is looking for and some expertise in manipulating it (see Chapter 8), then the options for improved GIS applications are greatly enhanced.

1.4 Global positioning systems

Usually, when we talk about remotely sensed data, we are referring to imagery – that is, a file that contains reflectance values for many points covering a given rectangular area. The global positioning system (**GPS**) is also based on satellite data, but the data consists of positions only – there is no attribute information other than some **metadata** on how the position was determined. Another difference is that GPS data can be collected on a continuing basis, which helps to collect not just single positions but also route data. In other words, while a remotely sensed image contains data about a lot of neighboring locations that gets updated on a daily to yearly basis, GPS data potentially consist of many irregularly spaced points that are separated by seconds or minutes.

As of 2006, there was only one easily accessible GPS world-wide. The Russian system as well as alternative military systems are out of reach of the typical GIS user, and the planned civilian European system will not be functional for a number of years. Depending on the type of receiver, ground conditions, and satellite constellations, the horizontal accuracy of GPS measurements lies between a few centimeters and a few hundred meters, which is sufficient for most GIS applications (however, buyer beware: it is never as good as vendors claim).

GPS data is mainly used to attach a position to field data – that is, to spatialize attribute measurements taken in the field. It is preferable for the two types of measurement to be taken concurrently because this decreases the opportunity for errors in matching measurements with their corresponding position. GPS data is increasingly augmented by a new version of triangulating one's position that is based on cellphone signals (Bryant 2005). Here, the three or more satellites are either replaced or preferably added to by cellphone towers. This increases the likelihood of having a continuous signal, especially in urban areas, where buildings might otherwise disrupt GPS reception. Real-time applications especially benefit from the ability to track moving objects this way.

1.5 Digitizing and scanning

Most spatial legacy data exists in the form of paper maps, sketches or aerial photographs. And although most newly acquired data comes in digital format, legacy data holds potentially enormous amounts of valuable information. The term *digitizing* is usually applied to the use of a special instrument that allows interactive tracing of the outline of **features** on an analogue medium (mostly paper maps). This is in contrast to **scanning**, where an instrument much like a photocopying or fax machine captures a digital image of the map, picture or sketch. The former creates geometries for geographic objects, while the latter results in a picture much like early uses of imagery to provide a backdrop for pertinent geometries.

Nowadays, the two techniques have merged in what is sometimes called on-screen or heads-up digitizing, where a scanned image is loaded into the GIS and the operator then traces the outline of objects of their choice on the screen. In any case, and parallel to the use of GPS measurements, the result is a file of mere geometries, which then have to be linked with the attribute data describing each geographic object. Outsiders keep being surprised how little the automatic recognition of objects has been advanced and hence how much labor is still involved in digitizing or scanning legacy data.

1.6 The attribute component of geographic data

Most of the discussion above concerns the geometric component of geographic information. This is because it is the geometric aspects that make spatial data special. Handling of the attributes is pretty much the same as for general-purpose data handling, say in a bank or a personnel department. Choice of the correct attribute, questions of classification, and error handling are all important topics; but, in most instances, a standard textbook on database management would provide an adequate introduction.

More interesting are concerns arising from the combination of attributes and geometries. In addition to the classical mismatch, we have to pay special attention to a particular geographic form of ecological fallacy. Spatial distributions are hardly ever uniform within a unit of interest, nor are they independent of scale.

2 Accessing Existing Data

Most GIS users will start using their systems by accessing data compiled either by the GIS vendor or by the organization for which they work. Introductory tutorials tend to gloss over the amount of work involved even if the data does not have to be created from scratch. Working with existing data starts with finding what's out there and what can be rearranged easily to fulfill one's data requirements. We are currently experiencing a sea change that comes under the buzz word of interoperability. GISystems and the data that they consist of used to be insular enterprises, where even if two parties were using the same software, the data had to exported to an exchange format. Nowadays different operating systems do not pose any serious challenge to data exchange any more, and with ubiquitous WWW access, the remaining issues are not so much technical in nature.

2.1 Data exchange

Following the logic of geographic data structure outlined in Chapter 1, data exchange has to deal with two dichotomies, the common (though not necessary) distinction between geometries and attributes, and the difference between the geographic data on the one hand and its cartographic representation on the other.

Let us have a closer look at the latter issue. Geographic data is stored as a combination of locational, attribute and possibly temporal components, where the locational part is represented by a reference to a virtual position or a boundary object. This locational part can be *represented* in many different ways – usually referred to as the mapping of a given geography. This mapping is often the result of a very laborious process of combining different types of geographic data, and if successful, tells us a lot more than the original tables that it is made up of (see Figure 5). Data exchange can then be seen as (1) the exchange of the original geography, (2) the exchange of only the map graphics – that is, the map symbols and their arrangement, or (3) the exchange of both. The translation from geography to map is a proprietary process, in addition to the user's decisions of how to represent a particular geographic phenomenon.

The first thirty years of GIS saw the exchange mainly of ASCII files in a proprietary but public format. These exchange files are the result of an export operation and have to be imported rather than directly read into the second system. Recent standardization efforts led to a slightly more sophisticated exchange format based on the Web's extensible markup language, **XML**. The ISO standards, however, cover only a minimum of commonality across the systems and many vendor-specific features are lost during the data exchange process.

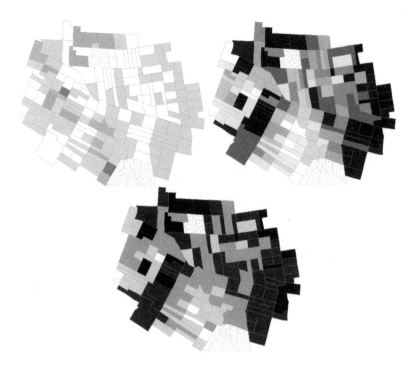

Figure 5 One geography but many different maps

2.2 Conversion

Data conversion is the more common way of incorporating data into one's GIS project. It comprises three different aspects that make it less straightforward than one might assume. Although there are literally hundreds of GIS vendors, each with their own proprietary way of storing spatial information, they all have ways of storing data using one of the de-facto standards for simple attributes and geometry. These used to be **dBASE**™ and **AutoCAD**™ exchange files but have now been replaced by the published formats of the main vendors for combined vector and attribute data, most prominently the ESRI shape file format, and the **GeoTIFF**™ format for pixel-based data. As there are hundreds of GIS products, the translation between two less common formats can be fraught with high information loss and this translation process has become a market of its own (see, for example, SAFE Corp's feature manipulation engine FME).

The second conversion aspect is more difficult to deal with. Each vendor, and arguably even more GIS users, have different ideas of what constitutes a geographic object. The translation of not just mere geometry but the **semantics** of what is encoded in a particular vendor's scheme is a hot research topic and has sparked a whole new branch of GIScience dealing with the ontologies of representing geography. A glimpse of the difficulties associated with translating between ontologies can be gathered from the differences between a raster and a vector representation of a geographic phenomenon. The academic discussion has gone beyond the raster/vector debate, but at the practical level this is still the cause of major headaches, which can be avoided only if all potential users of a GIS dataset are involved in the original definition of the database semantics. For example, the description of a specific shoal/sandbank depends on whether one looks at it as an obstacle (as depicted on a nautical chart) or as a seal habitat, which requires parts to be above water at all times but defines a wider **buffer** of no disturbance than is necessary for purely navigational purposes.

The third aspect has already been touched upon in the section on data exchange – the translation from geography to map data. In addition to the semantics of geographic features, a lot of effort goes into the organization of spatial data. How complex can individual objects be? Can different vector types be mixed, or vector and raster definitions of a feature? What about representations at multiple scales? Is the **projection** part of the geographic data or the map (see next section)? There are many ways to skin a cat. And these ways are virtually impossible to mirror in a conversion from one system to another. One solution is to give up on the exchange of the underlying geographic data and to use a desktop publishing or web-based **SVG** format to convert data from and to. These provide users with the opportunity to alter the graphical representation. The ubiquitous **PDF** format, on the other hand, is convenient because it allows the exchange of maps regardless of the recipient's output device but it is a dead end because it cannot be converted into meaningful map or geography data.

2.3 Metadata

All of the above options for conversion depend on a thorough documentation of the data to be exchanged or converted. This area has seen the greatest progress in recent years as **ISO** standard 19115 has been widely adopted across the world and across many disciplines (see Figure 6). A complete metadata specification of a geospatial dataset is extremely labor-intensive to compile and can be expected only for relatively new datasets, but many large private and government organizations mandate a proper documentation, which will eventually benefit the whole geospatial community.

2.4 Matching geometries (projection and coordinate systems)

There are two main reasons why geographic data cannot be adequately represented by simple geometries used in popular computer aided design (**CAD**) programs. The first is that projects covering more than a few square kilometers have to deal with the curvature of the Earth. If we want to depict something that is little under the horizon, then we need to come up with ways to flatten the earth to fit into our two-dimensional computer world. The other reason is that, even for smaller areas, where the curvature could be neglected, the need to combine data from different sources, especially satellite imagery – requires matching coordinates from different

Metadata
Identification Information
   Citation
   Description
   Time Period of Content
   Status
   Spatial Reference
      Horizontal Coordinate System Definition: planar
      Map Projection: Lambert conformal conic
      Standard parallel: 43.000000
      Standard parallel: 45.500000
      Longitude of Central Meridian: −120.500000
      Latitude of Projection Origin: 41.750000
      False Easting: 1312336.000000
      False Northing: 0.000000
      Abcissa Resolution: 0.004096
      Ordinate Resolution: 0.004096
      Horizontal Datum: NAD83
      Ellipsoid: GRS80
      Semi-major Axis: 6378137.000000
      Flattening Ratio: 298.572222
   Keywords
   Access Constraints
Reference Information
   Metadata Date
   Metadata Contact
   Metadata Standard Name
   Metadata Standard Version

Figure 6 Subset of a typical metadata tree

coordinate systems. The good news is that most GIS these days relieve us from the burden of translating between the hundreds of projections and coordinate systems. The bad news is that we still need to understand how this works to ask the right questions in case the metadata fails to report on these necessities.

Contrary to Dutch or Kansas experiences as well as the way we store data in a GIS, the Earth is not flat. Given that calculations in spherical geometry are very complicated, leading to rounding errors, and that we have thousands of calculations performed each time we ask the GIS to do something, manufacturers have decided to adopt the simple two-dimensional view of a paper map. Generations of cartographers have developed a myriad of ways to map positions on a sphere to coordinates on flat paper. Even the better of these projections all have some flaws and the main difference between projections is the kind of distortion that they introduce to the data (see Figure 7). It is, for example, impossible to design a map that measures the distances between all cities correctly. We can have a table that lists all these distances but there is no way to draw them properly on a two-dimensional surface.

Many novices to geographic data confuse the concepts of projections and coordinate systems. The former just describes the way we project points from a sphere on to a flat surface. The latter determines how we index positions and perform measurements on the result of the projection process. The confusion arises from the fact that many geographic

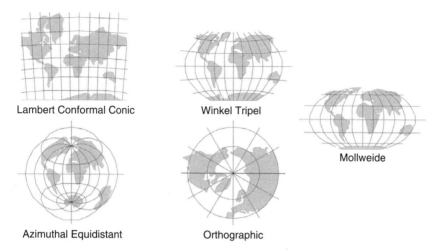

Lambert Conformal Conic   Winkel Tripel

Mollweide

Azimuthal Equidistant   Orthographic

Figure 7   The effect of different projections

coordinate systems consist of projections and a mathematical coordinate system, and that sometimes the same name is used for a geographic coordinate system and the projection(s) it is based on (e.g. the Universal Transverse Mercator or **UTM** system). In addition, geographic coordinate systems differ in their metric (do the numbers that make up a coordinate represent feet, meters or decimal degrees?), the definition of their origin, and the assumed shape of the Earth, also known as its geodetic datum. It goes beyond the scope of this book to explain all these concepts but the reader is invited to visit the USGS website at http://erg.usgs.gov/isb/pubs/factsheets/fs07701.html for more information on this subject.

Sometimes (e.g. when we try to incorporate old sketches or undocumented maps), we do not have the information that a GIS needs to match different datasets. In that case, we have to resort to a process known as rubber sheeting, where we interactively try to link as many individually identifiable points in both datasets to gain enough information to perform a geometric transformation. This assumes that we have one master dataset whose coordinates we trust and an unknown or untrusted dataset whose coordinates we try to improve.

2.5 Geographic web services

The previous sections describe a state of data acquisition, which is rapidly becoming outdated in some application areas. Among the first questions that one should ask oneself before embarking on a GIS project is how unique is this project? If it is not too specialized then chances are that there is a market for providing this service or at least the data for it. This is particularly pertinent in application areas where the geography changes constantly, such as a weather service, traffic monitoring, or real estate markets. Here it would be prohibitively expensive to constantly collect data for just one application and one should look for either data or if one is lucky even the analysis results on the web.

Web-based geographic data provision has come a long (and sometimes unexpected) way. In the 1990s and the first few years of the new millennium, the emphasis was on **FTP** servers and web portals that provided access to either public domain data (the USGS and US Census Bureau played a prominent role in the US) or to commercial data, most commonly imagery. Standardization efforts, especially those aimed at congruence with other IT standards, helped geographic services to become mainstream. Routing services (like it or not, **MapQuest** has become a household name for what geography is about), neighborhood searches such as local.yahoo.com, and **geodemographics** have helped to catapult geographic web services out of the academic realm and into the marketplace. There is an emerging market for non-GIS applications that are yet based on the provision of decentralized geodata in the widest sense. Many near real-time applications such as sending half a million volunteers on door-to-door canvassing during the 2004 presidential elections in the US, the forecast of avalanche risks and subsequent day-to-day operation of ski lifts in the European Alps, or the coordination of emergency management efforts during the 2004 tsunami have only been possible because of the interoperability of web services.

The majority of web services are commercial, accessible only for a fee (commercial providers might have special provisions in case of emergencies). As this is a very new market, the rates are fluctuating and negotiable but can be substantial if there are many (as in millions) individual queries. The biggest potential lies in the emergence of middle-tier applications not aimed at the end user that are based on raw data and transform these to be combined with other web services. Examples include concierge services that map attractions around hotels with continuously updated restaurant menus, department store sales, cinema schedules, etc., or a nature conservation website that continuously maps GPS locations of collared elephants in relationship to updated satellite imagery rendered in a 3-D landscape that changes according to the direction of the track. In some respect, this spells the demise of GIS as we know it because the tasks that one would usually perform in a GIS are now executed on a central server that combines individual services the same way that an end consumer used to combine GIS functions. Similar to the way that a **Unix** shell script programmer combines little programs to develop highly customized applications, web services application programmers now combine traditional GIS functionality with commercial services (like the one that performs a secure credit card transaction) to provide highly specialized functionality at a fraction of the price of a GIS installation.

This form of outsourcing can have great economical benefits and, as in the case of emergency applications, may be the only way to compile crucial information at short notice. But it comes at the price of losing control over how data is combined. The next chapter will deal with this issue of quality control in some detail.

3 Handling Uncertainty

The only way to justifiably be confident about the data one is working with is to collect all the primary data oneself and to have complete control over all aspects of acquisition and processing. In the light of the costs involved in creating or accessing existing data this is not a realistic proposition for most readers.

GIS own their right of existence to their use in a larger spatial decision-making process. By basing our decisions on GIS data and procedures, we put faith in the truthfulness of the data and the appropriateness of the procedures. Practical experience has tested that faith often enough for the GIS community to come up with ways and means to handle the uncertainty associated with data and procedures over which we do not have complete control. This chapter will introduce aspects of spatial data quality and then discuss metadata management as the best method to deal with spatial data quality.

3.1 Spatial data quality

Quality, in very general terms, is a relative concept. Nothing is or has innate quality; rather quality is related to purpose. Even the best weather map is pretty useless for navigation/orientation purposes. Spatial data quality is therefore described along characterizing dimensions such as positional **accuracy** or thematic **precision**. Other dimensions are completeness, consistency, **lineage**, semantics and time.

One of the most often misinterpreted concepts is that of accuracy, which often is seen as synonymous with quality although it is only a not overly significant part of it. Accuracy is the inverse of error, or in other words the difference between what is supposed to be encoded and what actually is encoded. 'Supposed to be encoded' means that accuracy is measured relative to the world model of the person compiling the data; which, as discussed above, is dependent on the purpose. Knowing for what purpose data has been collected is therefore crucial in estimating data quality. This notion of accuracy can now be applied to the positional, the temporal and the attribute components of geographic data. Spatial accuracy, in turn, can be applied to points, as well as to the connections between points that we use to depict lines and boundaries of area features. Given the number of points that are used in a typical GIS database, the determination of spatial accuracy itself can be the basis for a dissertation in spatial statistics. The same reasoning applies to the temporal component of geographic data. Temporal accuracy would then describe how close the recorded time for a crime event, for instance, is to when that crime actually took place. Thematic accuracy, finally, deals with how close the match is between the attribute value that should be there and that which has been encoded. For quantitative measures this is determined similarly to positional accuracy. For qualitative measures, such as the correct land use classification of a pixel in a remotely sensed image, an error classification matrix is used.

Precision, on the other hand, refers to the amount of detail that can be discerned in the spatial, temporal or thematic aspects of geographic information. Data modelers prefer the term 'resolution' as it avoids a term that is often confused with accuracy. Precision is indirectly related to accuracy because it determines to a degree the world model against which the accuracy is measured. The database with the lower precision automatically also has lower accuracy demands that are easier to fulfill. For example, one land use categorization might just distinguish commercial versus residential, transport and green space, while another distinguishes different kinds of residential (single-family, small rental, large condominium) or commercial uses (markets, repair facilities, manufacturing, power production). Assigning the correct thematic association to each pixel or feature is considerably more difficult in the second case and in many instances not necessary. Determining the accuracy and precision requirements is part of the thought process that should precede every data model design, which in turn is the first step in building a GIS database.

Accuracy and precision are the two most commonly described dimensions of data quality. Probably next in order of importance is database consistency. In traditional databases, this is accomplished by normalizing the tables, whereas in geographic databases **topology** is used to enforce spatial and temporal consistency. The classical example is a cadastre of property boundaries. No two properties should overlap. Topological rules are used to enforce this commonsense requirement; in this case the rule that all two-dimensional objects must intersect at one-dimensional objects. Similarly, one can use topology to ascertain that no two events take place at the same time at the same location. Historically, the discovery of the value of topological rules for GIS database design can hardly be overestimated.

Next in order of commonly sought data quality characteristics is completeness. It can be applied to the conceptual model as well as to its implementation. Data model completeness is a matter of mental rigor at the beginning of a GIS project. How do we know that we have captured all the relevant aspects of our project? A stakeholder meeting might be the best answer to that problem. Particularly on the implementation side, we have to deal with a surprising characteristic of completeness referred to as over-completeness. We speak of an error of commission when data is stored that should not be there because it is outside the spatial, temporal or thematic bounds of the specification.

Important information can be gleaned from the lineage of a dataset. Lineage describes where the data originally comes from and what transformations it has gone through. Though a more indirect measure than the previously described aspects of data quality, it sometimes helps us make better sense of a dataset than accuracy figures that are measured against an unknown or unrealistic model.

One of the difficulties with measuring data quality is that it is by definition relative to the world model and that it is very difficult to unambiguously describe one's world model. This is the realm of semantics and has, as described in the previous chapter, initiated a whole new branch of information science trying to unambiguously describe all relevant aspects of a world model. So far, these **ontology** description languages are able to handle only static representations, which is clearly a shortcoming where even GIS are now moving into the realm of process orientation.

3.2 How to handle data quality issues

Many jurisdictions now require mandatory data quality reports when transferring data. Individual and agency reputations need to be protected, particularly when geographic information is used to support administrative decisions subject to appeal. On the private market, firms need to safeguard against possible litigation by those who allege to have suffered harm through the use of products that were of insufficient quality to meet their needs. Finally, there is the basic scientific requirement of being able to describe how close information is to the truth it represents.

The scientific community has developed formal models of uncertainty that help us to understand how uncertainty propagates through spatial processing and decision-making. The difficulty lies in communicating uncertainty to different levels of users in less abstract ways. There is no one-size-fits-all to assess the fitness for use of geographic information and reduce uncertainty to manageable levels for any given application. In a first step it is necessary to convey to users that uncertainty is present in geographic information as it is in their everyday lives, and to provide strategies that help to absorb that uncertainty.

In applying the strategy, consideration has initially to be given to the type of application, the nature of the decision to be made and the degree to which system outputs are utilized within the decision-making process. Ideally, this prior knowledge permits an assessment of the final product quality specifications to be made before a project is undertaken; however, this may have to be decided later when the level of uncertainty becomes known. Data, software, hardware and spatial processes are combined to provide the necessary information products. Assuming that uncertainty in a product is able to be detected and modeled, the next consideration is how the various uncertainties may best be communicated to the user. Finally, the user must decide what product quality is acceptable for the application and whether the uncertainty present is appropriate for the given task.

There are two choices available here: either reject the product as unsuitable and select uncertainty reduction techniques to create a more accurate product, or absorb (accept) the uncertainty present and use the product for its intended purpose.

In summary, the description of data quality is a lot more than the mere portrayal of errors. A thorough account of data quality has the chance to be as exhaustive as the data itself. Combining all the aspects of data quality in one or more reports is referred to as metadata (see Chapter 2).

4   Spatial Search

Among the most elementary database operations is the quest to find a data item in a database. Regular databases typically use an indexing scheme that works like a library catalog. We might search for an item alphabetically by author, by title or by subject. A modern alternative to this are the indexes built by desktop or Internet search engines, which basically are very big lookup tables for data that is physically distributed all over the place.

Spatial search works somewhat differently from that. One reason is that a spatial coordinate consists of two indices at the same time, x and y. This is like looking for author and title at the same time. The second reason is that most people, when they look for a location, do not refer to it by its x/y coordinate. We therefore have to translate between a spatial reference and the way it is stored in a GIS database. Finally, we often describe the place that we are after indirectly, such as when looking for all dry cleaners within a city to check for the use of a certain chemical.

In the following we will look at spatial queries, starting with some very basic examples and ending with rather complex queries that actually require some spatial analysis before they can be answered. This chapter does deliberately omit any discussion of special indexing methods, which would be of interest to a computer scientist but perhaps not to the intended audience of this book.

4.1 Simple spatial querying

When we open a spatial dataset in a GIS, the default view on the data is to see it displayed like a map (see Figure 8). Even the most basic systems then allow you to use a query tool to point to an individual feature and retrieve its attributes. They key word here is 'feature'; that is, we are looking at databases that actually store features rather than field data.

If the database is raster-based, then we have different options, depending on the sophistication of the system. Let's have a more detailed look at the right part of Figure 8. What is displayed here is an elevation dataset. The visual representation suggests that we have contour lines but this does not necessarily mean that this is the way the data is actually stored and can hence be queried by. If it is indeed line data, then the current cursor position would give us nothing because there is no information stored for anything in between the lines. If the data is stored as areas (each plateau of equal elevation forming one area), then we could move around between any two lines and would always get the same elevation value. Only once we cross a line would we 'jump' to the next higher or lower plateau. Finally, the data could be

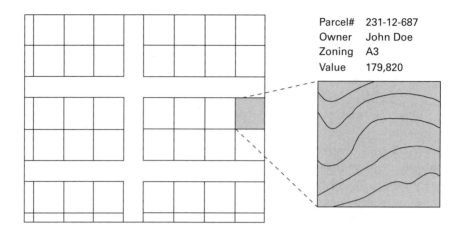

Parcel#   231-12-687
Owner   John Doe
Zoning   A3
Value   179,820

Figure 8   Simple query by location

stored as a raster dataset, but rather than representing thousands of different elevation values by as many colors, we may make life easier for the computer as well as
for us (interpreting the color values) by displaying similar elevation values with only
one out of say 16 different color values. In this case, the hovering cursor could still
query the underlying pixel and give us the more detailed information that we could
not possibly distinguish by the hue.

This example illustrates another crucial aspect of GIS: the way we store data has
a major impact on what information can be retrieved. We will revisit this theme
repeatedly throughout the book. Basically, data that is not stored, like the area
between lines, cannot simply be queried. It would require rather sophisticated analytical techniques to interpolate between the lines to come up with a guesstimate for
the elevation when the cursor is between the lines. If, on the other hand, the elevation is explicitly stored for every location on the screen, then the spatial query is
nothing but a simple lookup.

4.2 Conditional querying

Conditional queries are just one notch up on the level of complication. Within a GIS,
the condition can be either attribute- or geometry-based. To keep it simple and get
the idea across, let's for now look at attributes only (see Figure 9).

Here, we have a typical excerpt from an attribute table with multiple variables. A
conditional query works like a filter that initially accesses the whole database.
Similar to the way we search for a URL in an Internet search engine, we now provide the system with all the criteria that have to be fulfilled for us to be interested in
the final presentation of records. Basically, what we are doing is to reject ever more
records until we end up with a manageable number of them. If our query is "Select
the best property that is >40,000m², does not belong to Silma, has tax code 'B', and

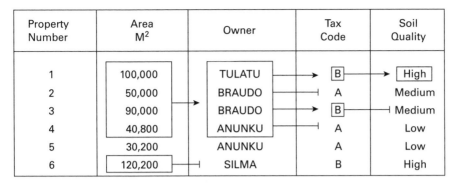

Property Number	Area M²	Owner	Tax Code	Soil Quality
1	100,000	TULATU	B	High
2	50,000	BRAUDO	A	Medium
3	90,000	BRAUDO	B	Medium
4	40,800	ANUNKU	A	Low
5	30,200	ANUNKU	A	Low
6	120,200	SILMA	B	High

Figure 9   Conditional query or query by (multiple) attributes

has soils of high quality", then we first exclude record #5 because it does not fulfill the first criterion. Our selection set, after this first step, contains all records but #5. Next, we exclude record #6 because our query specified that we do not want this owner. In the third step, we reduce the number of candidates to two because only records #1 and #3 survived up to here and fulfill the third criterion. In the fourth step, we are down to just one record, which may now be presented to us either in a window listing all its attributes or by being highlighted on the map.

Keep in mind that this is a pedagogical example. In a real case, we might end up with any number of final records, including zero. In that case, our query was overly restrictive. It depends on the actual application, whether this is something we can live with or not, and therefore whether we should alter the query. Also, this conditional query is fairly elementary in the way it is phrased. If the GIS database is more than just a simple table, then the appropriate way to query the database may be to use one dialect or another of the structured query language **SQL**.

4.3 The query process

One of the true benefits of a GIS is that we have a choice whether we want to use a tabular or a map interface for our query. We can even mix and match as part of the query process. As this book is process-oriented, let's have a look at the individual steps. This is particularly important as we are dealing increasingly often with Internet GIS user interfaces, which are difficult to navigate if the sequence and the various options on the way are not well understood (see Figure 10).

First, we have to make sure that the data we want to query is actually available. Usually, there is some table of contents window or a legend that tells us about the data layers currently loaded. Then, depending on the system, we may have to select the one data layer we want to query. If we want to find out about soil conditions and the 'roads' layer is *active* (the terminology may vary a little bit), then our query result will be empty. Now we have to decide whether we want to use the map or the tabular interface. In the first instance, we pan around the map and use the identify

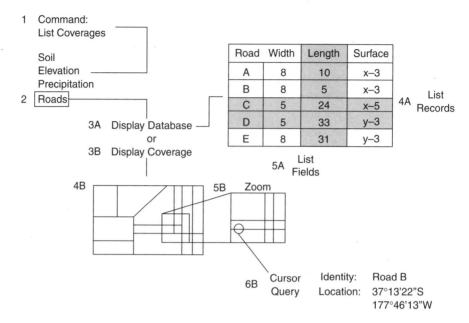

Figure 10   The relationship between spatial and attribute query

tool to learn about different restaurants around the hotel we are staying at. In the second case, we may want to specify '*Thai cuisine under $40*' to filter the display. Finally, we may follow the second approach and then make our final decision based on the visual display of what other features of interest are near the two or three restaurants depicted.

4.4 Selection

Most of the above examples ended with us selecting one or more records for subsequent manipulation or analysis. This is where we move from simple mapping systems to true GIS. Even the selection process, though, comes at different levels of sophistication. Let's look at Figure 11 for an easy and a complicated example.

In the left part of the figure, our graphical selection neatly encompasses three features. In this case, there is no ambiguity – the records for the three features are displayed and we can embark on performing our calculations with respect to combined purchase price or whatever. On the right, our selection area overlaps only partly with two of the features. The question now is: do we treat the two features as if they got fully selected or do we work with only those parts that fall within our search area? If it is the latter, then we have to perform some additional calculations that we will encounter in the following two chapters.

One aspect that we have glanced over in the above example is that we actually used one geometry to select some other geometries. Figure 12 is a further illustration of

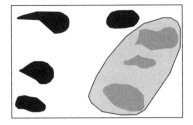 

Stand	Name	Area	Species
A–3	North	20	Pine
C–2	East–1	10	Pine
C–2	East–2	40	Mix

Stand	Name	Area	Species
A–3	North	10	Pine
C–2	East–1	5	Pine
C–2	East–2	30	Mix

Figure 11   Partial and complete selection of features

the principle. Here, we use a subset of areas (e.g. census areas) to select a subset of point features such as hospitals. What looks fairly simple on the screen actually requires quite a number of calculations beneath the surface. We will revisit the topic in the next chapter.

4.5 Background material: Boolean logic

This topic is not GIS-specific but is necessary background for the next two chapters. Those who know **Boolean logic** may merrily jump to the next chapter, the others should have a sincere look at the following.

Boolean logic was invented by English mathematician George Bool (1815–64) and underlies almost all our work with computers. Most of us have encountered Boolean logic in queries using Internet search engines. In essence, his logic can be described as the kind of mathematics that we can do if we have nothing but zeros and ones. What made him so famous (after he died) was the simplicity of the rules to combine those zeros and ones and their powerfulness once they are combined. The basic three operators in Boolean logic are NOT, OR and AND.

Figure 13 illustrates the effect of the three operators. Let's assume we have two GIS layers, one depicting income and the other depicting literacy. Also assume that the two variables can be in one of two states only, high or low. Then each location can be a combination of high or low income with high or low literacy. Now we can look at Figure 13. On the left side we have one particular spatial configuration – not all that realistic because it's not usual to have population data in equally sized spatial units, but it makes it a lot easier to understand the principle. For each area, we can read the values of the two variables.

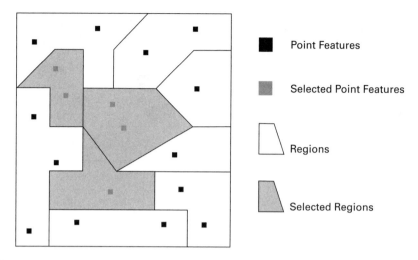

Figure 12 Using one set of features to select another set

HL HI	HL LI	LL HI
LL LI	HL HI	HL LI
LL HI	LL LI	HI HI

HL: High Literacy
LL: Low Literacy
HI: High Income
LI: Low Income

HL HI	**HL LI**	LL HI
LL LI	HL HI	**HL LI**
LL HI	**LL LI**	HI HI

Not HI

HL HI	**HL LI**	**LL HI**
LL LI	**HL HI**	**HL LI**
LL HI	**LL LI**	**HI HI**

HL or HI

HL HI	**HL LI**	LL HI
LL LI	HL HI	**HL LI**
LL HI	LL LI	HI HI

HL not HI

HL HI	HL LI	LL HI
LL LI	**HL HI**	HL LI
LL HI	LL LI	**HI HI**

HL and HI

Figure 13 Simple Boolean logic operations

Now we can query our database and, depending on our use of Boolean operators, we gain very different insights. In the right half of the figure, we see the results of four different queries (we get to even more than four different possible outcomes by combining two or more operations). In the first instance, we don't query about literacy at all. All we want to make sure is that we reject areas of high income, which leaves us with the four highlighted areas. The NOT operator is a unary operator – it affects only the descriptor directly after the operand, in this first instance the income layer.

Next, look at the OR operand. Translated into plain English, OR means 'one or the other, I don't care which one'. This is in effect an easy-going operand, where only one of the two conditions needs to be fulfilled, and if both are true then the better. So, no matter whether we look at income or literacy, as long as either one (or both) is high, the area gets selected. OR operations always result in a maximum number of items to be selected.

Somewhat contrary to the way the word is used in everyday English, AND does not give us the combination of two criteria but only those records that fulfill *both* conditions. So in our case, only those areas that have both high literacy and high income at the same time are selected. In effect, the AND operand acts like a strong filter. We saw this above in the section on conditional queries, where all conditions had to be fulfilled.

The last example illustrates that we can combine Boolean operations. Here we look for all areas that have a high literacy rate but not high income. It is a combination of our first example (NOT *HI*) with the AND operand. The result becomes clear if we rearrange the query to state NOT *HI* AND *HL*. We say that AND and OR are binary operands, which means they require one descriptor on the left and one on the right side. As in regular algebra, parentheses () can be used to specify the sequence in which the statement should be interpreted. If there are no parentheses, then NOT precedes (overrides) the other two.

5   Spatial Relationships

Spatial relationships are one of the main reasons why one would want to use a GIS. Many of the cartographic characteristics of a GIS can be implemented with a drawing program, while the repository function of large spatial databases is often taken care of by traditional database management systems. It is the explicit storage of spatial relationships and/or their analysis based on geometric reasoning that distinguishes GIS from the rest of the pack.

We ended the last chapter with a select-by-location operation, which already makes use of a derived relationship between areas and points that lay either inside or outside these areas. Before we embark on a discussion of many other important spatial relationships, we should insert a little interlude in the form of the spatial database operation 'recode'. Functionally, and from the perspective of typical GIS usage, this operation sits in between simple spatial queries and more advanced analytical functions that result in new data.

5.1 Recoding

Recoding is an operation that is usually applied when the contents of a database have become confusingly complicated; as such it is used to simplify (our view of) the database. Soil maps, such as the one depicted in Figure 14, are a perfect example of that. Ten different soil types may be of interest to the **pedologist**, but for most others it is sufficient to know whether the ground is stable enough to build a high-rise or dense enough to prevent groundwater leakages. In that case, we would like to aggregate the highly detailed information contained in a soils database and recoding is the way to do it.

Figure 15 is a stylized version of the previous soil map and illustrates how the combination of attributes also leads to a combination of geometries. We will make use of this side effect in the next chapter, when, as a result of combining spatial data, we have more geometries than we would like. Alternatively, we could use the recoding operation not as much to simplify our view of the database but to reflect a particular interpretation of the data. A simple application of this is given in Figure 16, where we simplify a complex map to a binary suitable/non-suitable for agricultural purposes.

A more complicated (and interesting) version of the same procedure is given in Figure 17. Here, we recode a complex landcover map by first extracting all different vegetation types and then recombine these to form a new dataset containing all kinds of vegetation and nothing but vegetation. In both of these examples, we are creating

Figure 14 Typical soil map

Typical soil map

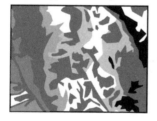

Simplified map with 10
different soil types

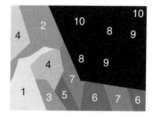

Attribute Code	Soil Type
1	A1Z
2	A3X
3	A2Z
4	A1H
5	B1H
6	B3X
7	B1X
8	C3H
9	C2H
10	C2X

Recoded soil map

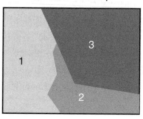

Attribute Recode	New Soil
1	A
2	B
3	C

Figure 15 Recoding as simplification

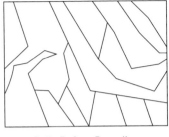

Soils Before Recoding

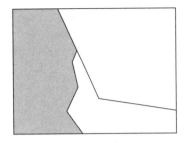

Agricultural Soils

Recode:
Agricultural Soils – A
Non-agricultural Soils – B, C

Figure 16   Recoding as a filter operation

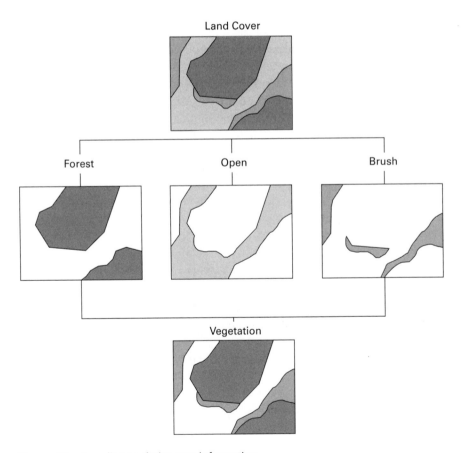

Figure 17   Recoding to derive new information

new data based on a new interpretation of already existing data. There is no category 'vegetation' in the original landcover dataset. We will revisit the topic of creating new data rather than just querying existing data in the next chapter on combining data. This sneak preview is an indicator of the split personality of the recoding operation; it could be interpreted as a mere data maintenance operation or as an analytical one.

One intriguing aspect that comes to mind when one looks at Figures 16 and 17 is that we immediately try to discern patterns in the distribution of selected areas. Spatial relationships can be studied quantitatively or qualitatively. The former will be the subject of Chapter 10, while the latter is addressed in the following sections. Features are defined by their boundaries. On the qualitative side, we can therefore distinguish between two types of spatial relationship, one where we look at how individual coordinates are combined to form feature boundaries and the other where we look at the spatial relationships among features.

5.2 Relationships between measurements

As discussed in Chapter 2, all locational references can be reduced to one or more coordinates, which are either measured or interpolated. It is important to remind ourselves that we are talking about the data in our geographic databases, not the geometries that are used to visualize the geographic data, which may be the same but most likely are not. If you are unsure about this topic, please revisit Chapter 2.

Next, we need to distinguish between the object-centered and the field-based representations of geographic information (see also Chapter 1). The latter does not have any feature representation, so the spatial relationships are reduced to those of the respective positions of pixels to each other. This then is very straightforward, as we have only a very limited number of scenarios, as depicted in Figure 18:

- Cell boundaries can touch each other.
- Cell corners can touch each other.
- Cells don't touch each other at all.
- Cells relate to each other not within a layer but across (vertically).

We will revisit the cell relationships in Chapter 8, when we look at the analytical capabilities of raster GIS – which are entirely based on the simplicity of their spatial relationships.

Features, on the other hand, are defined by their boundaries. We distinguish zero- through 3-dimensional simple features from their complex counterparts (see Figure 19). One and the same node, edge or area can be shared by any number of higher-dimensional features. Older GIS consisted of tables of points, lines and areas, which all consisted of pointers to the respective lower-dimensional tables (see Figure 20). Modern systems store the nodes of higher-dimensional features redundantly and use topological rules (see next section) to enforce database integrity. Complex features consist of multiple disconnected geometries that are treated as one uniform object.

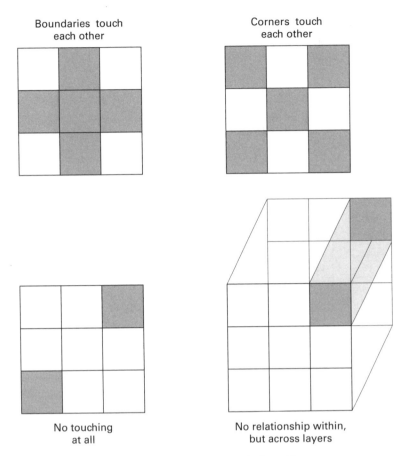

Figure 18   Four possible spatial relationships in a pixel world

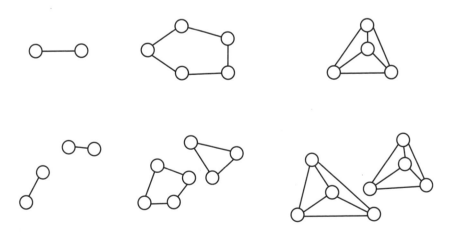

Figure 19   Simple (top row) and complex (bottom row) geometries

Node table		Line table			Area table	
ID	x,y	ID	from to		ID	lines
01	07,30	01	01,02		01	1,2,17,16,14,15
02	12,35	02	02,04		02	3,4,5,19,18,17
03	15,22	03	04,07		03	6,7,8,9,10,21,20,18,19
04	20,37	04	07,11		04	20,21,11,12,13,16
05	25,20	05	11,12			
06	28,30	06	12,16			
07	27,42	07	16,18			
08	28,15	08	18,17			
09	32,14	09	17,15			
10	32,30	10	15,14			
11	32,40	11	14,09			
12	35,35	12	09,08			
13	35,22	13	08,05			
14	35,15	14	05,03			
15	42,20	15	03,01			
16	42,37	16	05,06			
17	50,28	17	06,04			
18	52,35	18	06,10			
		19	10,12			
		20	10,13			
		21	13,14			

Figure 20   Pointer structure between tables of feature geometries

Networks form an extra category of relationships between measurements. Similar to one-dimensional features, the connections between any two nodes are more abstract. In the world of features, we assume a spatial relationship, when two elements have the same coordinate. In networks, the default is no relationship, unless it has been explicitly stored as a node attribute. A network is understood as a graph whose lines may or may not intersect even when they visually do. The technical term for this is non-planarity, which means that although we draw the network on a flat plane, the intersecting lines may actually mean to be at different levels. A subway map would be a practical example of that; if the lines intersect without a station symbol (a node) then there is no connectivity, which in turn means that passengers cannot switch from one line to the other (see Figure 21). The qualitative spatial relationship is hence one of connectedness – whether a node is reachable or not.

5.3 Relationships between features

The qualitative spatial relationship between features is again described in a form of connectedness called topology. Topology is the branch of mathematics that deals with persistence of relationships, no matter how much we contort the objects of

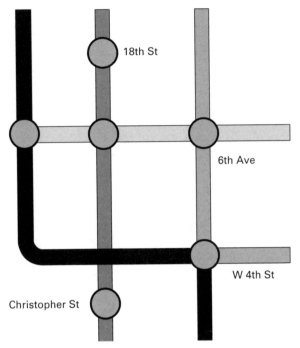

Figure 21   Part of the New York subway system

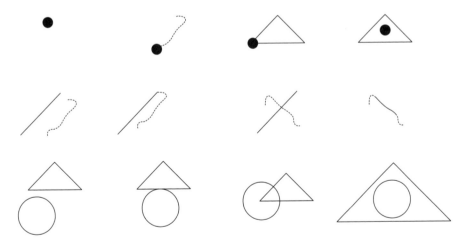

Figure 22   Topological relationships between features

interest, short of cutting them up. We distinguish between containment, intersecting, touching, and no relationship whatsoever, and do so for the relationships across a range of dimensionalities. We hence have all possible combinations of the above relationships among and between points, lines, areas and volumes (Egenhofer 1993).

Figure 22 lists a selection of such topological relationships. Volumes have been written about what are the mathematically distinguishable qualitative relationships, how they relate to quantitative measures, and whether we even have names for them in different human languages. In English, for instance, we use verbs such as touch, border, intersect, overlap and contain, to describe the relationships depicted in Figure 22.

The use of topological rules has proven to be a very useful instrument for checking the consistency of a GIS database. Until recently, GIS could store only one type of geometry per feature class. Spatial relationships within one and the same feature class can be encoded as part of the data structure, which imposes additional constraints on the data organization and helps to check for database consistency. If these rules are applied across feature classes, then we have to (a) perform a topology check by at least temporarily combining spatial features, and (b) store these in an extra table that is associated with a group of feature classes. We will revisit this topic in the next chapter about combining spatial data.

6 Combining Spatial Data

As mentioned in the previous chapter, many spatial relationships are difficult to derive or even describe. Rather than storing all possible relationships between all features of a database, we can use GIS operations to answer specific questions about the spatial relationships among our features of interest. This is not only far more efficient but also gives us more freedom, because we can determine on the fly what pieces of data we want to relate to each other – and in the age of the Internet, these pieces may even be distributed across the world.

This chapter deals with two families of GIS operations that in practice make up some 75–80% of all analytical GIS operations. *Overlay* is the quintessential GIS operation that seems to define a GIS. If a software package can perform overlay operations, then it is indeed a GIS and not a mere CAD or cartography program. The poor cousin is the *buffer* operation, which always seems to be mentioned second. Both are actually place-holders for a number of different operations, but we will discuss this in detail in the following.

6.1 Overlay

In the recoding section of the last chapter, we saw how tightly linked attributes and geometries are. By recombining attributes we automatically changed the graphical representation as well. However, when we look at it from the perspective of lines store in our database (see Figure 20), the recoding operation did not create any *new* geometries. This changes now with the group of overlay operations. Let's look at Figure 23 to see what happens to geometries and attributes in an overlay operation.

For pedagogical reasons, we use very simple geometries and only two layers of binary data. This is extremely unrealistic but helps us to get the principle across. We will look at more realistic examples later on.

The figure shows a number of important aspects of the overlay operation. We have two or more input feature classes and one (new) output feature class. The geometries of the two input layers are most likely to be different; that is, they do not have to come from the same provider and do not have to have anything in common other than the general extent (it does not make sense to overlay data in South America with other data from Africa). Also observe that the first feature class has no attributes describing vegetation, while the second has none describing soils. The overlay operation depicted here looks for coincidences at the same location. In other words, for any given location, it looks what information there is in one feature class, then in the

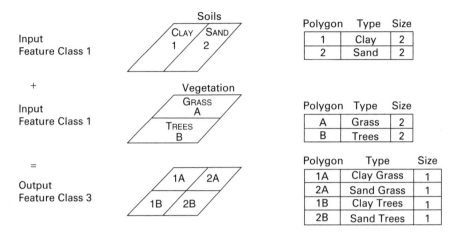

Polygon	Type	Size
1	Clay	2
2	Sand	2

Polygon	Type	Size
A	Grass	2
B	Trees	2

Polygon	Type	Size
1A	Clay Grass	1
2A	Sand Grass	1
1B	Clay Trees	1
2B	Sand Trees	1

Figure 23 Schematics of a polygon overlay operation

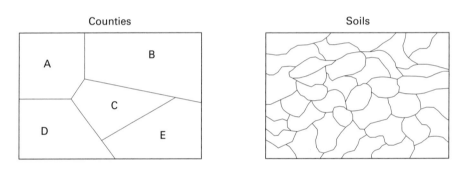

Which soils occur in County C?

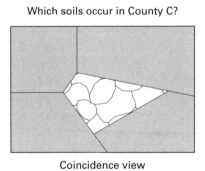

Coincidence view

Figure 24 Overlay as a coincidence function

other feature class, and then it combines the two in the output dataset. The geometries of the output dataset did not exist before. We really create completely new data.

We used this notion of coincidence earlier when we looked at some of the more advanced spatial search functions. Figure 24 is a case in point.

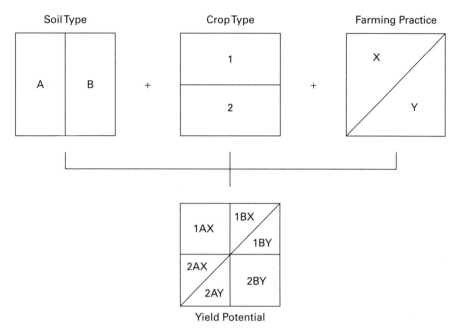

Figure 25   Overlay with multiple input layers

The question 'Which soils occur in county C?' appears to be a simple spatial search function. However, what actually happens here is that we overlay the two feature classes and then look for those parts of the soils feature class that fall within the polygon that marks county C in the counties feature class. In Chapter 4, we did not have an answer to how to deal with those soil polygons that lie only partially within county C. Now, we applied the overlay operation like a cookie cutter and created several new pieces of data. When you look back at Figure 23, you see that there is quite a lot happening here. New files or at least tables get created, new geometries have to be calculated, new connections between attributes and geometries have to be formed and maintained. This may seem like a lot for a simple spatial search operation – and it is! Keep this in mind, the next time you wait impatiently for the result of your spatial search.

What we did in Figure 24 was to overlay one layer of areal features with another one. A slightly more complicated (realistic) example is depicted in Figure 25. Here we are using three input layers and possibly some weighting scheme to calculate agricultural yield potential. Going beyond the pure area-on-area situation, we could just as well overlay areal with point or line feature classes, for example to determine which historic site is in what administrative unit or to ascertain all the countries that the river Nile is flowing through. We could even overlay point and line feature classes to learn whether the Christopher St subway station is on the red or the blue line (see Figure 21). In each of these cases we make use of the topological relationships between all the features involved.

6.2 Spatial Boolean logic

In Chapter 4, we looked briefly at Boolean logic as the foundation for general computing. You may recall that the three basic Boolean operators were NOT, AND and OR. In Chapter 4, we used them to form query strings to retrieve records from attribute tables. The same operators are also applicable to the combination of geometries; and in the same way that the use of these operators resulted in very different outputs, the application of NOT, AND and OR has completely different effects on the combination (or overlay) of layer geometries.

Figure 26 illustrates the effect of the different operands in a single overlay operation. This is why we referred to overlay as a group of functions. Figure 26 is possibly the most important in this book. It is not entirely easy to digest the information provided here and the reader is invited to spend some time studying each of the situations depicted. Again, for pedagogical reasons, there are only two layers with only one feature each. In reality, the calculations are repeated thousands of times when we overlay two geographic datasets. What is depicted here is the resulting geometry only. As in the example of Figure 23 above, all the attributes from all the input layers are passed on to the output layer.

Depending on whether we use one or two Boolean operators and how we relate them to the operands, we get six very different outcomes. Clearly one overlay is not the same as the other. At the risk of sounding overbearing, this really is a very important figure to study. GIS analysis is dependent on the user understanding what is

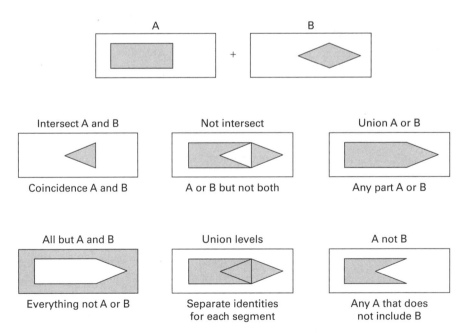

Figure 26 Spatial Boolean logic

happening here and being able to instruct whatever system is employed to perform the correct overlay operation.

The relative success of the overlay operations can be attributed to their cognitive consonance with the way we detect spatial patterns. Overlays are instrumental in answering questions like 'What else can be observed at this location?', or 'How often do we find woods and bison at the same place?'.

6.3 Buffers

Compared to overlay, the buffer operation is more quantitative if not analytical. And while, at least in a raster-based system, we could conceive of overlay as a pure database operation, buffering is as spatial as it gets. Typically, a buffer operation creates a new area around our object of interest – although we will see exotic exceptions from this rule. The buffer operation takes two parameters: a buffer distance and the object around which the buffer is to be created. The result can be observed in Figure 27.

A classical, though not GIS-based, example of a buffer operation can be found in every larger furniture store. You will invariably find some stylized or real topographic map with concentric rings usually drawn with a felt pen that center on the location of the store or their storage facility. The rings mark the price that the store charges for the delivery of their furniture. It is crude but surprisingly functional.

Regardless of the dimension of the input feature class (point, line or polygon), the result of a regular buffer operation is always an area. Sample applications for points would be no-fly zones around nuclear power plants, and for lines noise buffers around highways. The buffer distance is usually applied to the outer boundary of the object to be buffered. If features are closer to each other than the buffer distance between them, then the newly created buffer areas merge – as can be seen for the two right-most groups of points in Figure 27.

There are a few interesting exceptions to the general idea of buffers. One is the notion of inward buffers, which by its nature can only be applied to one- or higher-dimensional features. A practical example would be to define the core of an ecological

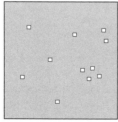 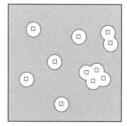 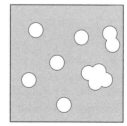

Original Points   Buffered Points   Dissolved Buffers

Figure 27   The buffer operation in principle

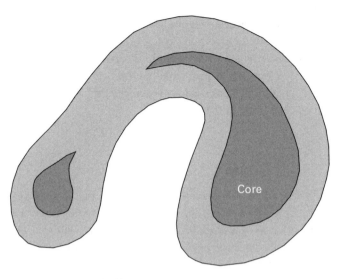

Figure 28   Inward or inverse buffer

Figure 29   Corridor function

reserve (see Figure 28). A combination of the regular and the inverse buffer applied simultaneously to all features of interest is called a corridor function (see Figure 29). Finally, within a street network, the buffer operation can be applied along the edges (a one-dimensional buffer) rather than the often applied but useless as-the-crow-flies circular buffer. We will revisit this in the next chapter.

6.4 Buffering in spatial search

A few paragraphs above we saw how overlay underlies some of the (not overtly) more complicated spatial search operations. The same holds true for buffering. Conceptually, buffers are in this case used as a form of neighborhood. 'Find all customers within ZIP code 123' is an overlay operation, but 'Find all customers in a radius of 5 miles' is a buffer operation. Buffers are often used as an intermediate select, where we use the result of the buffer operation in subsequent analysis (see next section).

6.5 Combining operations

If the above statement that buffers and overlays make up in practice some 75% of all analytical GIS functionality is true, then how is it that GIS has become such an important genre of software? The solution to this paradox lies in the fact that operations can be concatenated to form workflows. The following is an example from a major flood in Mozambique in 2000 (see Figure 30).

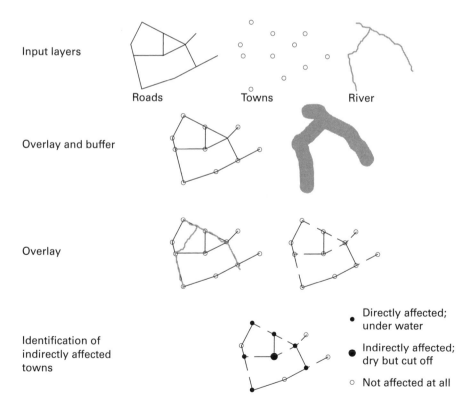

Figure 30   Surprise effects of buffering affecting towns outside a flood zone

We start out with three input layers – towns, roads and hydrology. The first step is to buffer the hydrology layer to identify flood zones (this makes sense only in coastal plains, such as was the case with the Southern African floods in 2000). Step two is to overlay the township layer with the flood layer to identify those towns that are directly affected. Parallel to this, an overlay of the roads layer with the flood layer selects those roads that have become impassable. A final overlay of the impassable roads layer with the towns helps us to identify the towns that are *indirectly* affected – that is, not flooded but cut off because none of the roads to these towns is passable. Figure 30 is only a small subset of the area that was affected in 2000.

6.6 Thiessen polygons

A special form of buffer is hidden behind a function that is called a **Thiessen polygon** (pronounced the German way as 'ee') or Voronoi diagram. Originally, these functions had been developed in the context of graph theory and applied to GIS based on triangulated irregular networks (**TIN**s), which we will discuss in Chapter 9. It is introduced here as a buffer operation because conceptually what happens is that each of the points of the input layer is simultaneously buffered with ever-increasing buffer size. Wherever the buffers hit upon each other, a 'cease line' is created until no buffer can increase any more. The result is depicted in Figure 31.

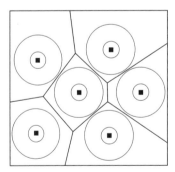 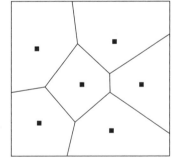

Figure 31   Thiessen polygons

Each location within the newly created areas is closer to the originating point than to any other one. This makes Thiessen polygons an ideal tool for allocation studies, which we will study in detail in the next chapter.

7   Location–Allocation

Among the main reasons for wanting to use a GIS are (1) finding a location, (2) finding the best way to get to that location, (3) finding the best location to do whatever our business is, and (4) optimizing the use of our limited resources to conduct our business. The first question has been answered at varying levels of complexity in the earlier chapters. Now I want to address the other three questions.

General GIS textbooks usually direct the reader to answer these questions by using the third and so far neglected form of GIS data structure, the network GIS. This is, however, slightly misleading as we could just as well use map algebra (Chapter 8), and some of the more advanced regional science models would even use data aggregated to polygons (although here the shape of the polygons and hence much of the reason why we would use vector GIS is not considered). The following notes are more about concepts; the actual procedures in raster or in network GIS would differ considerably from each other. But that is an implementation issue and should not be of immediate concern to the end user.

7.1 The best way

Finding the best way to a particular location is usually referred to as shortest-path analysis. But that is shorthand for a larger group of operations, which we will look at here. To determine the best way one needs at a minimum an origin and a desti-nation. On a featureless flat plain, the direct line between these two locations would mark the best way. In the real world, though, we have geography interfering with this simple geometric view. Even if we limit ourselves to just the shortest distance, we tend to stay on streets (where available), don't walk through walls, and don't want to get stuck in a traffic jam. Often, we have other criteria but pure distance that determine which route we choose: familiarity, scenery, opportunity to get some other business done on the way, and so on. Finally, we typically are not the only ones to embark on a journey, say from home to work. Our decisions, our choice of what is the best way, are influenced by what other people are doing, and they are time-dependent. An optimal route in the morning may not easily be traced back in the evening. In most general terms, what we are trying to accomplish with our best-way analysis is to model the flows of commodities, people, capital or information over space (Reggiani 2001). How, then, can all these issues be addressed in a GIS, and how does all this get implemented?

A beginning is to describe the origin and the target. This could be done in the form of two coordinate pairs, or a relative position given by distance and direction from an origin. Either location can be imbued with resources in the widest sense, possibly better described as push and pull factors. Assuming for a moment that the origin is a point (node, **centroid**, pixel), we can run a wide range of calculations on the attributes of that point to determine what factors make the target more desirable than our origin and what resources to use to get there. The same is true for any point in between that we might visit or want to avoid. Finally, we have to decide how we want to travel. There *may* be a constraining geometry underlying our geography. In the field view perspective we could investigate all locations within our view shed, whereas in a network we would be constrained by the links between the nodes. These links usually have a set of attributes of their own, determining speed, capacity (remember, we are unlikely to be the only ones with the wish to travel), or mode of transport. In a raster GIS, the attributes for links and nodes are combined at each pixel, which actually makes it easier to deal with hybrid functionality such as turns. Turn tables are a special class of attribute table that permit or prevent us from changing direction; they can also be used to switch modes of transportation. Each pixel, node or link could have its own schedule or a link to a big central time table that determines the local behavior at any given time in the modeling scenario.

The task is then to determine the best way among all the options outlined above. Two coordinate pairs and a straight line between them rarely describes our real world problem adequately (we would not need a GIS for that). The full implementation of all of the above options is as of writing this book just being tested for a few mid-sized cities. Just to assemble all the data (before even embarking on developing the routing algorithms) is a major challenge. Given the large number of options, we are faced with an optimization problem. The implementation is usually based on graph theoretical constructs (forward star search, Dijkstra algorithm) and will not be covered here. But conceptually, the relationship between origins and targets is based on the gravity model, which we will look at in the following section.

7.2 Gravity model

In the above section, we referred to the resources that we have available and talked about the push and pull of every point. This vocabulary is borrowed from a naive model of physics going all the way back to Isaac Newton. Locations influence each other in a similar way that planets do in a solar system. Each variable exerts a field of influence around its center and that field is modeled using the same equations that were employed in mechanics. This intellectual source has provided lots of ammunition for social scientists who thought the analogy to be too crude. But modern applications of the gravity model in location–allocation models are as similar to Newton's role model as a GPS receiver to a compass.

The gravity model in spatial analysis is the inductive formalization of Tobler's **First Law** (see Chapter 10). Mathematically, we refer to a distance–decay function, which in Newton's case was one over the square of distance but in spatial analysis can be a wide range of functions. By way of example, $2 may get me 50 km away from the central station in New York, 20 km in Hamburg, Germany, and nowhere in Detroit if my mode of transport is a subway train. We can now associate fields of influence based on a number of different metrics with each location in our dataset (see Figure 32). Sometimes they act as a resource as in our fare example, sometimes they act as an attractor that determines how far we are willing to access a certain resource (school, hospital, etc.). Sometimes they may even act as a distracter, an area that we don't want to get too close to (nuclear power plants, prisons, predators).

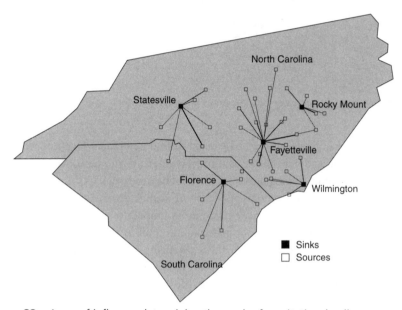

Figure 32 Areas of influence determining the reach of gravitational pull

This push and pull across all known locations of a study area forms the basis for answering the next question, finding the optimal location or site for a particular resource, be it a new fire station or a coffee shop. The next section will describe the concepts behind location modeling.

7.3 Location modeling

Finding an optimal location has been the goal of much research in business schools and can be traced all the way back to nineteenth and early twentieth century scholars such as von Thünen, Weber and Christaller. The idea of the gravity model applies to all of them (see Figures 33–35), albeit in increasingly complicated ways. Von Thünen worked on an isolated agricultural town. Weber postulated a simple triangle of resource, manufacturer and market location. Christaller expanded this view into a whole network of spheres of influence.

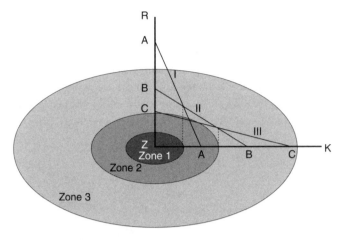

Figure 33 Von Thünen's agricultural zones around a market

In the previous chapter, if we had wanted to find an optimal location, we would have used a combination of buffer and overlay operations to derive the set of locations, whose attribute combination and spatial characteristics fulfill a chosen criterion. While the buffer operation lends a bit of spatial optimization, the procedure (common as it is as a pedagogical example) is limited to static representations of territorial characteristics. Location modeling has a more human-centered approach and captures flows rather than static attributes, making it much more interesting. It tries to mimic human decision choices at every known location (node, cell or area). Weber's triangle (Figure 34) is particularly illustrative of the dynamic character of the weights pulling our target over space.

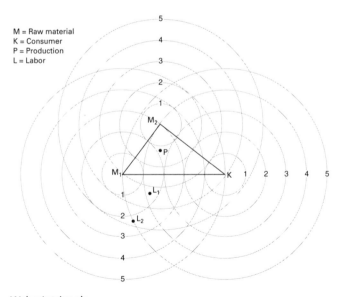

Figure 34 Weber's triangle

Two additions to this image drive the analogy home. Rather than having a plane surface, we model the weights pulling our optimal center across some rugged terrain. Each hill and peak marks push factors or locations we want to avoid. The number of weights is equivalent to the number of locations that we assume to have an influence over our optimal target site. The weights themselves finally consist of as many criteria given as much weight as we wish to apply. The weights could even vary depending on time of day, or season, or real-time sensor readings. The latter would then be an example for the placement of sentinels in a public safety scenario.

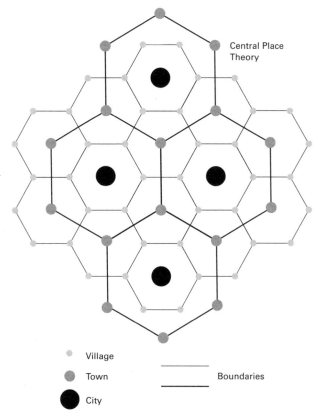

Central Place
Theory

Village

Town

City

Boundaries

Figure 35   Christaller's Central Place theory

The implementation of such a system of gravity models is fairly straightforward for a raster model (as we will see in the discussion of zonal operations in the following chapter) or a network model (particularly if our commodities are shipped along given routes). For a system of regions interacting with each other, the implementation is traditionally less feature-based. Instead, large input–output tables representing the flows from each area to each other area are used in what is called a flow matrix (see Figure 36). The geometry of each of these areas is neglected and the flows are aggregated to one in each direction across a boundary. Traditionally employed in regional science applications, the complications of geometry are

To Destinations

From	Zone 1	Zone 2	Zone 3	Row sums
Zone 1	27	4	16	47
Zone 2	9	23	4	36
Zone 3	0	6	20	26
Column sums	36	33	40	109

Figure 36   Origin-destination matrix

overridden by the large number of variables (weights) that are pulling our target cell across the matrix.

7.4 Allocation modeling

All of the above so far assumed that there is only one target location that we either want to reach or place. If this decision has already been made (by us and/or our competitors) then the question arises as to what is the next best location. As in the statistical urn game, we may want to pursue this question with or without the option of moving already existing sites. And finally, we may want to find out when the rate of diminishing returns means that we have saturated the market (the term 'market' is here to be seen in a very wide sense; we could talk about placement of policemen, expensive instruments, any non-ubiquitous item). Allocation models are the domain of optimization theory and operations research, and the spatial sciences have not made many inroads into these fields. In the course of this chapter, the problems tackled and the required toolset have grown ever bigger. Allocation models, if they are supposed to show any resemblance with reality, are enormously complicated and require huge amounts of data – which often does not exist (Alonso 1978). The methods discussed in Chapter 11, in particular a combination of genetic algorithms, **neural networks** and agent-based modeling systems, may be employed to address these questions in the future.

The discussion above illustrates how models quickly become very complicated when we try to deal with a point, line and polygon representation of geographic phenomena. Modelers in the natural sciences did not abandon the notion of space to the degree that regional scientists do and turned their back on spatial entities rather than space itself. In other words, they embraced the field perspective, which is computationally a lot simpler and gave them the freedom to develop a plethora of advanced spatial modeling tools, which we will discuss in the next chapter.

8   Map Algebra

This chapter introduces the most powerful analytical toolset that we have in GIS. Map algebra is inherently raster-based and therefore not often taught in introductory GIS courses, except for applications in resource management. Traditional vector-based GIS basically knows the buffer and overlay operations we encountered in Chapter 6. The few systems that can handle network data then add the location–allocation functionality we encountered in Chapter 7. All of that pales in comparison to the possibilities provided by map algebra, and this chapter can really only give an introduction. Please check out the list of suggested readings at the end of this chapter.

Map Algebra was invented by a chap called Dana Tomlin as part of his PhD thesis. He published his thesis in 1990 under the very unfortunate title of *Cartographic Modeling* and both names are used synonymously. His book (Tomlin 1990) deserves all the accolades that it received, but the title is really misleading, as the techniques compiled in it have little if anything to do with cartography.

The term 'map algebra' is apt because it describes arithmetic on cells, groups of cells, or whole feature classes in form of equations. Every map algebra expression has the form <output = *function*(input)>. The function can be unary (applying to only one operand), binary (combining two operands as in the elementary arithmetic functions plus, minus, multiply and divide), or *n*-ary, that is applying to many operands at once.

We distinguish map algebra operations by their spatial scope; local functions operate on one cell at a time, neighborhood functions apply to cells in the immediate vicinity, zonal functions apply to all cells of the same value, and global functions apply to all cells of a layer/feature class. In spite of the scope, all map algebra functions work on a cell-by-cell basis. The scope only determines how many other cells the function takes into consideration, while calculating the output value for the cell it currently operates on (see Figure 37). However, before we get into the details of map algebra functions, we have to have a look at how raster GIS data is organized.

8.1 Raster GIS

Raster datasets can come in many disguises. Images – raw, georeferenced, or even classified – consist of raster data. So do many thematic maps if they come from a natural resource environment, digital elevation models (see Chapter 9), and most dynamic models in GIS. As you may recall from Chapter 2, a raster dataset describes the location and characteristics of an area and their relative position in space. A single raster dataset typically describes a single theme such as land use or elevation.

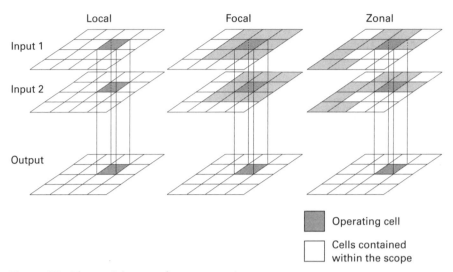

Figure 37 The spatial scope of raster operations

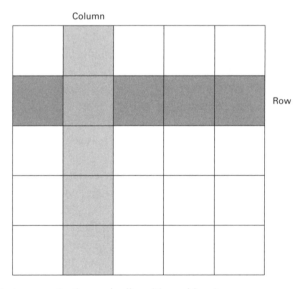

Figure 38 Raster organization and cell position addressing

At the core of the raster dataset is the cell. Cells are organized in rows and columns and have a cell value – very much like spreadsheets (see Figure 38). To prove this point, Waldo Tobler, in a 1992 article, described building a GIS using Microsoft Excel; you are not encouraged to follow that example as the coding of GIS functionality is extremely cumbersome and definitely not efficient. Borrowing from the nomenclature of map algebra, all cells of the same value are said to belong to the same zone (see Figure 39). Cells that are empty – that is, for which there is no known value – are marked as NoData. NoData is different from 0 (zero) or –9999, or any

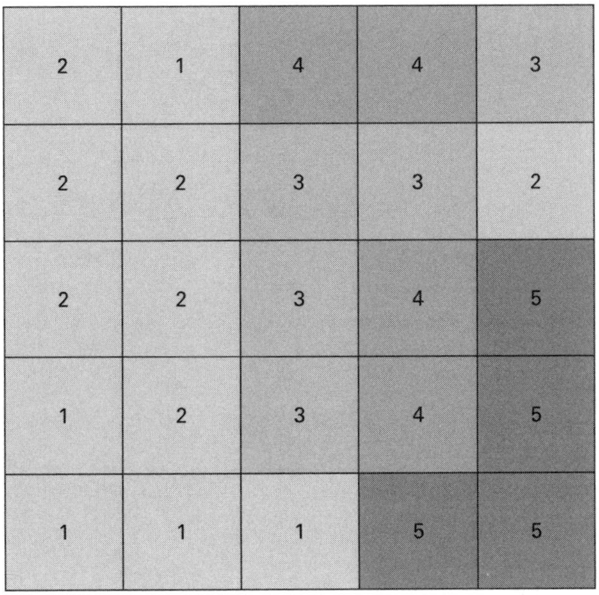

Figure 39   Zones of raster cells

other typically out-of-range value. Upon encountering a NoData cell, map algebra functions react in a well-defined way.

Cell values can have two different purposes. They can represent a true quantitative value (e.g. elevation or amount of precipitation), or they can represent a class, whose values are then described in an external table. In the latter case, the cell value acts as a pointer to the correct record in the external table.

8.2 Local functions

All map algebra functions work one cell at a time. Local functions derive their name from the fact that in the calculation of the output value only input cells with exactly the same coordinate are considered (see Figure 40). The somewhat tedious description of the procedure goes as follows.

A local map algebra function reads the cell values of cell position (1,1) and applies a certain calculation on these input values. It then writes the result to cell (1,1) in the output layer and proceeds to the second cell in the row, where the whole procedure is repeated, one cell at a time until we get to the last cell in the last row. It is easy to see that this would be very slow if each reading/writing step were to involve the hard disk. With the price of memory coming down, most GIS nowadays are able to process one or two layers (depending on their size) virtually before writing the result to a file or table. The cumbersomeness of the process is mitigated by the fact that no complicated geometric calculations have to be performed (as all cells

Local

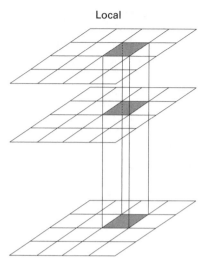

Figure 40   Local function

are of the same size and orientation), and that the calculations themselves can often be executed within the processor itself, which makes it extremely fast. So, although a million cells may be processed, the result of a local operation is often instantaneous. Compare this to the complexity of overlay operations in the vector world (see Chapter 6)!

Local map algebra functions can be arithmetic, trigonometric, exponential, logarithmic, statistical or logical in nature. A trivial example would involve only one input layer, where we multiply all cell values by a constant value, say '3'. As a result, we could have an elevation layer, where the ratio of horizontal to vertical distances is now exaggerated enough to visually discern terrain features (see Figure 41).

One notch up on the ladder of sophistication is the use of a multiplier layer (see Figure 42). Say we have counties with different property tax rates. We could then calculate not just purchase prices but long-term costs by multiplying the costs of

2	0	1	1
2	3	0	4
4		2	3
1	1		2

× 3 =

6	0	3	3
6	9	0	12
12		6	9
3	3		6

Figure 41   Multiplication of a raster layer by a scalar

2	0	1	1
2	3	0	4
4		2	3
1	1		2

×

6	0	3	3
6	9	0	12
12		6	9
3	3		6

=

12	0	3	3
12	27	0	48
48		12	27
3	3		12

Figure 42   Multiplying one layer by another one

each property by the tax rates. Observe what happens to the cells with no values; NoData times something results in NoData.

8.3 Focal functions

Neighborhood or **focal functions** are the main reason for the success of map algebra. Local functions, in a way, are nothing but fancy recoding operations, which could indeed easily be performed in a proper database management system. Focal functions, on the other hand, take into account all cells in a user-defined neighborhood. Anybody who ever tried to calculate a cell value in a spreadsheet based on a combination of surrounding cells knows that the query string very quickly becomes really complicated. Again, the procedure is one cell at a time, except that for the calculation of an output cell value, we now look at all cells surrounding the processing location of all input layers (see Figure 43).

Focal

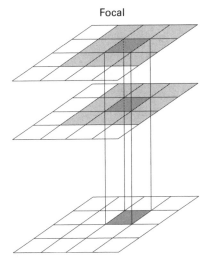

Figure 43   Focal function

By default, the neighborhood is defined to be all eight cells touching the processing cell, plus the processing cell in the middle. As an interesting aside, the first step of a focal function uses only three neighboring cells because cell (1,1) sits in a corner and has no more neighbors. Many GIS allow us to define different neighborhoods, though. We may change the extent (size) of the neighborhood, its shape, or its kernel (where in relationship to the neighborhood sits the operating cell?).

Just about any function type (arithmetic, trigonometric, exponential, logarithmic, statistical or logical) that we encountered for local functions is also applicable for focal functions. In addition, we have functions that compare cell values within a neighborhood, rather than across multiple input layers. This way, we can determine the average, minimum or maximum value within the neighborhood or calculate the span (range) of values. One of the attractive features of map algebra is its naming conventions. Identifiers such as FocalSum, FocalMean, LocalAND or ZonalMax are more or less self-explanatory, yet this is exactly how all map algebra functions are formed. Actual vendor implementations change a little bit in their semantics (e.g. average versus mean) but stick to the principle.

Figure 44 describes a prominent focal function that is the basis for many image processing operations. You are encouraged to work the example, or at least the first row of cells in Figure 44. Remember to include the processing cell in all calculations. We will revisit neighborhood functions in Chapter 9, as the very notion of terrain is a focal one.

2	0	1	1
2	3	0	4
4	2	2	3
1	1	3	2

=

1.8	1.3	1.5	1.5
2.2	2.0	1.8	1.8
2.2	2.0	2.2	2.3
2.0	2.2	2.2	2.5

Figure 44 Averaging neighborhood function

8.4 Zonal functions

Zonal functions are in effect a mixture of local and focal functions. Based on Tobler's **First Law of Geography** (see Chapter 10), cells of similar values can be expected to lie next to each other. Hence everything that was said about focal functions applies to zonal functions as well. On the other hand, more important than spatial contiguity is the fact that all cells within a zone have the same value and they are treated the same. Zonal functions therefore take on the character of recoding functions that typified local functions.

The process is a bit more complicated than before. Again, zonal functions work on a cell-by-cell basis. However, the zone now acts as a kind of variable neighborhood definition (see Figure 45), and the zonal input layer as a lookup table. A zonal function processes at least two input layers, one so-called zone layer, and one or more value layers. The zones can be (and usually are) but do not have to be contiguous.

Zonal

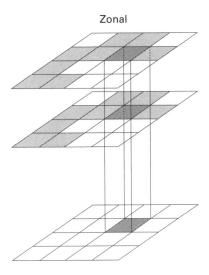

Figure 45   Zonal function

Figure 46 is an abstract example for a ZonalMax function. Again, as all map algebra functions, it starts at cell (1,1), and looks what zonal number it contains. Then it looks at all other cells with the same zonal number and creates a list of their values from the second input layer. Upon determining the maximum value, it writes that value to cell (1,1) in the output layer and proceeds to the next cell. If that next cell has the same zonal number then the output value from the previous step is written to cell (1,2). If the zonal number is different, then a new list of values belonging to all cells of the new zone number is created and its maximum is written to cell (1,2). In practice, the zonal number grids represent classified images or thematic maps, where we run calculations on the value layer based on the cell's membership to one class or another.

8.5 Global functions

Global functions differ from the rest in that many of the arithmetic or statistical operations do not make sense if applied globally (they would result in the same value for all cells of the output grid). Instead, global functions determine the spatial relationship between cells of interest – or in other words, the distance between cells.

Zone grid

2	2	1	1
2	3	3	1
	3	2	
1	1	2	2

Value grid

1	2	3	4
5	6	7	8
1	2	3	4
5	5	5	5

=

5	5	8	8
5	7	8	7
	5	8	
8	8	5	5

Figure 46   Value grids as spatial lookup tables

There are two types of global or distance function, Euclidean and weighted. The former is a straightforward geometric calculation, while the latter involves a second input layer that encodes a cost or friction surface. With global functions, it is fairly easy to implement the most complicated vector GIS operations: Thiessen polygons, corridors and location–allocation.

8.6 Map algebra scripts

One of the most convincing arguments for map algebra was Tomlin's use of map algebra scripts, a concatenation of functions that represents a complete GIS analysis workflow or even a dynamic model. While Tomlin just used simple sequences, authors like Kirby and Pazner (1990), Wesseling and van Deursen (1995) and Pullar (2003) developed temporal constructs that mimic classic programming languages augmented by spatial expressions. A simple example is the following script, which models the spread of a pollutant from a point source:

```
For i = 1 to 100
plume = buffer(plume) * 1/i
```

One of the interesting characteristics of most raster-based programs (particularly image processing packages) is that the target file can be overwritten in a programming loop. This saves disk space but of course prevents the analysis of intermediate steps. Scripts are also the mechanism behind the raster-based terrain modeling operations discussed in Chapter 9.

9   Terrain Modeling

All GIS functionality discussed so far assumed the world to be flat. This is, of course, utterly unrealistic but GIS has succeeded in getting away with this deficiency surprisingly well. Since 3-D representations are commonplace in modern CAD software, the obvious question is why has GIS not followed suit? The answer lies in the distinction between mere data representation that is common to CAD and vector GIS, on the one hand, and the spatial reasoning capacity that is unique to GIS.

In general, we can distinguish between three different ways to represent the third dimension (see Figure 47). One is the use of true three-dimensional coordinates to construct what in CAD is called a wire frame model. The second uses the same coordinates but constructs a mesh of triangles to drape a surface going through all the points. Applied to relatively small areas but at high precision, this data structure is often used in civil engineering applications. The last uses raster cells and records for each cell an elevation value as the pertinent attribute. The image of the raster surface in Figure 47 indicates a fourth representation, the use of contour lines. They are a filter on the former, leaving only values at certain intervals, which are then connected by a polyline. As we have seen earlier, the way we organize data directly determines what kind of questions we can ask of the data. The wire frame model is up to now the most limited of 3-D GIS models, serving mostly rendering functions similar to those of architectural software packages.

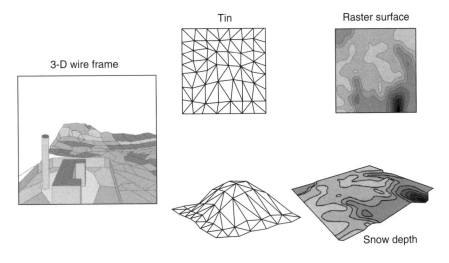

Figure 47   Three ways to represent the third dimension

9.1 Triangulated irregular networks (TINs)

While the wire frame data at least in theory allows for a truly three-dimensional representation (i.e., multiple objects intersecting with each other in all three dimensions), triangulated irregular networks or TINs are used for surfaces only. This may look like a limitation but it is not because none of the readily available off-the-shelf GIS have extended their analytical capacities to the third dimension. TINs are functionally a hybrid between the precise point representations typical for vector data and the field view of raster data; as a matter of fact, we could interpret them as an agglomeration of tiny fields, each represented by a triangular area. The triangles satisfy the Delaunay criterion, which ensures that no node lies within the interior of any of the circumcircles of the triangles (see Figure 48). This assures that the minimum interior angle of all of the triangles is maximized and long, thin triangles are avoided. In the process of calculating the best fitting triangles between all the points that make up a given surface, the GIS determines and stores slope and aspect of each. This is useful for two reasons. For one, this information is used in the rendering of 3-D surfaces and speeds up the process of displaying intricately shaded terrain. Second, it makes it easy to calculate volumes of irregularly shaped bodies – a task of particular pertinence in civil engineering.

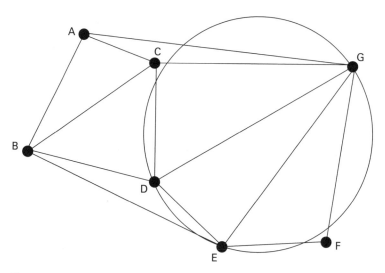

Figure 48   Construction of a TIN

TINs and raster-based elevation models are often contrasted as incorporating the classic difference between raster and vector GIS – the former being simple but either imprecise or huge; the latter being complicated, precise and very efficient. In practice, the precision of TINs is a function of how precise the original measurement was – which can be surprisingly poor in the third dimension. Also, most useful TINs consist of millions of points, either because of how they were derived from raster datasets, or because the surfaces used for civil engineering are so varied and have to reflect many sharp edges that we would not encounter in natural settings. The main difference then boils down to TINs being used (as the name implies) where we have irregularly spaced measurements, and raster-based elevation models being regular.

9.2 Visibility analysis

Viewshed analysis was originally a classical raster GIS function, typically applied to forest clear cuts in an attempt to determine which areas are visible from a road and should hence be left to avoid the wrath of environmentalists (Bettinger and Sessions 2003) (see Figure 49).

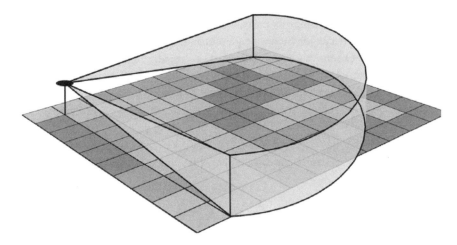

Figure 49   Viewshed

When cellphone usage took off in the 1990s, providers looked for tools that would help them to maximize coverage and minimize efforts for optimal placement of cellphone towers. The challenge for these companies was twofold. On the one hand they had to provide nationwide coverage as fast as possible and digital elevation models (see next section) are the obvious choice for that. On the other hand, the majority of customers are located in cities with a very different and impeding topography, which is better represented by TINs that are derived from high-resolution imagery.

Cellphone signals travel the same way as light does, and more important are obstructed in a similar way. In a raster environment, we are hence looking at a repeated zonal buffer operation. Due to the different data structure, the algorithms for a TIN work very differently; they are based on the same computational geometry that film studios now use for rendering.

9.3 Digital elevation and terrain models

Raster-based digital elevation models (DEMs) are by far the most common basis for dealing with the third dimension. As a matter of fact, many people have difficulties conceptualizing a raster attribute as anything but elevation. This is especially true for digital elevation models that cover larger areas. There are two reasons for this. For one, automated data capture for large areas tends to involve remote sensing methods that result in raster datasets. As important, however, is the fact that map algebra offers a plethora of analytical methods that require nothing but elevation and result in an astounding array of derivatives.

In Chapter 8, we saw how focal or neighborhood operations are used to analyze the combined spatial and attribute relationship between locations. The first set of focal operations transforms a DEM to a digital terrain model or DTM. The difference between the two is that the latter stores, in addition to mere elevation, two important derivatives – slope and aspect (see Figure 50). Slope is the ratio of elevation difference between the higher neighbors and the lower neighbors, and horizontal distance between those neighbors. It is usually calculated based on the 8-neighborhood, which in turn defines the distance as twice the cell width in the cardinal directions, and $\sqrt{2}$ times the former distance for the diagonals. Once we know the slope at each cell location, it is another simple focal operation to compare all neighbors and determine in what direction the surface slopes; this is called *aspect*. Given that both operations are fairly simple, and that processing power of modern computers is sufficient to calculate slope and aspect efficiently, DTMs are decreasing in importance.

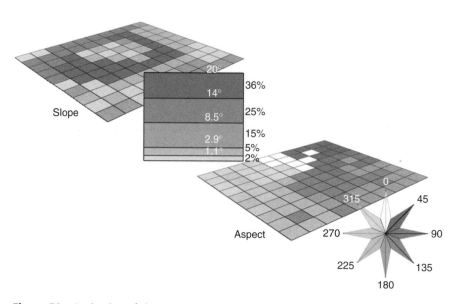

Figure 50   Derivation of slope and aspect

9.4 Hydrological modeling

Slope and aspect are the cornerstones of hydrological modeling. The underlying principle is fairly simple – and sometimes the application of out-of-the-box hydrological GIS functions is a bit simplistic. The idea is that once we know slope and aspect, we have a fairly good idea in what direction a drop of water would run off once it hits the surface. Clearly, in reality this is applicable only in either barren terrain or on totally saturated soils. On the other hand, all these complications can be (and are) addressed by more complicated map algebra scripts that combine infiltration and runoff models.

To get back to the basic idea, once we know into which neighboring cell a drop of water would flow, we can derive a flow direction map for the whole study area, which in turn becomes the basis for a flow accumulation map (see Figure 51). Basically, what happens here is that we look at each cell from the point of view of how many cells it receives from, and into which cell it passes its accumulated flow. We are virtually draining the terrain, and as a result see where streams would flow (Wilson & Gallant 2000). As mentioned above, this simple perspective works best in 'well-behaving' terrain, where there is a discernable slope and the land cover is not too variable. Under those conditions the resulting runoff calculation is surprisingly accurate; and with ever-increasing precision of DEM data, a range of national and international organizations is currently developing sophisticated hydrological models. The work at the University of Texas' Center for Research in Water Resources is fairly representative for these endeavors; their efforts to provide access to all their models and data via their website at http://www.crwr.utexas.edu/ is exemplary.

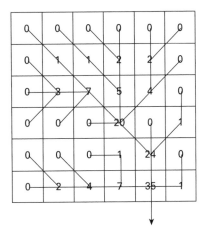

Figure 51   Flow accumulation map

Two aspects of hydrological modeling deserve extra mention. As can be seen from these four sections, all we needed in the beginning was a DEM. Each of the maps was a direct result of applying one operation or script to its predecessor. It is therefore fairly straightforward to write a rather longish script that creates all the intermediate files and derives a flood forecast model in one go – a prime example for the power of map algebra. Second, impressive as all of this is, it provides only a hint at all the models that can be developed and linked with each other. For example, based on a comparison of satellite images, we can develop models of urban sprawl and impervious surfaces. These can then be linked with the runoff model above to develop true simulation models, possibly linked with real-time readings of precipitation.

All of the above are true terrain modeling applications. The methods discussed in this chapter are, however, just as applicable to artificial surfaces, where the third dimension represents an attribute value such as property value or air pollution. These kinds of surfaces are typically interpolated from sparse measurements. The following chapter will cover spatial interpolation techniques and more.

10 Spatial Statistics

This chapter consists of two distinct parts because the phrase 'spatial statistics' has two rather different meanings, depending where one comes from. The first comes from a geophysics background and is also known as geo-statistics. The second has its roots in analytical cartography and is known as spatial analysis.

10.1 Geo-statistics

Typically, stochastic methods are applied when we have lots of data but not a really good idea of what drives the data. Deterministic models, on the other hand, are used when the data is sparse but we are confident that our understanding of the phenomenon, say based on a long history of similar observations, is good. The following set of methods cover both approaches, although the underlying commonality is a scarcity of data. All spatial interpolation techniques discussed here aim at creating surfaces from point data. It is important to recognize that most applications are not in terrain modeling but that the surfaces are intended to allow estimates of any kind of continuous variable, such as pollution, crime or mineral resources.

All model-based interpolation methods assume Tobler's (1970) *First Law of Geography*, which basically means that everything is related to everything else, but things that are closer in space (i.e., neighbors) are also closer with respect to their attribute values. There are many interpolation methods and we use some descriptors to categorize them by the number of input values (local versus global methods) and the characteristics of the resulting surface. If the surface goes through the values of known locations, then we label a method as exact. Otherwise we refer to it as a smooth interpolator, because the resulting surface avoids ruggedness around measured locations. The most common of these methods are inverse distance weighting (**IDW**), polynomials and splines.

10.1.1 Inverse distance weighting

IDW is the most direct implementation of Tobler's First Law (see Figure 52). Given known values at some locations, we try to determine the most probable value at a location for which we don't have a measurement.

The calculation is made up of known values of neighboring locations, which are weighted according to their distance from the one location for which we want to determine a value. This calculation is repeated for as many points as we care about. If the surface to be created is based on a TIN data structure, then we probably only double or triple the number of original points. If the surface is a raster dataset, then

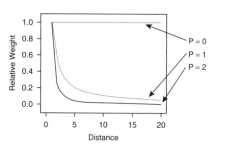
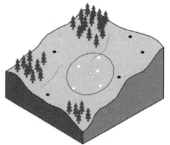

Figure 52 Inverse distance weighting

the calculation is repeated for every cell for which we don't have a measurement. The implementation of IDW differs among software packages, but most of them allow specification of the number and or distance of known values to be included, and in order to function properly they must allow for the user to specify the rate at which a location's weight decreases over distance. The differences lie in how sophisticated that distance–decay function can be. Because IDW calculates new values only for points for which no measurements exist, it does not touch the values of known locations and hence is an exact interpolator.

10.1.2 Global and local polynomials

Most readers will remember polynomials from their high school geometry classes. These are equations that we use to fit a line or curve through a number of known points. We encountered them in their simplest form in the calculation of slope, usually described in the form $y = a + bx$. Here we fit a straight line between two points, which works perfectly well in a raster GIS, where the distance from one elevation value to the next is minimal.

If the distance between the measured point locations is large, however, then a straight line is unlikely to adequately represent the surface; it would also be highly unusual for all the measured points to line up along a straight line (see Figure 53). Polynomials of second or higher degree (the number of plus or minus signs in the equation determines the degree of a polynomial) represent the actual surface much better.

Increasingly higher degrees have two disadvantages. First, the math to solve higher degree polynomials is quite complicated (remember your geometry class?). Second, even more importantly, a very sophisticated equation is likely to be an overfit. An overfit occurs when the equation is made to fit one particular set of input points but gets thrown off when that set changes or even when just one other point is added. In practice, polynomials of second or third degree have proven to strike the best balance.

We distinguish between so-called local and global polynomials, depending on whether we attempt to derive a surface for all our data or for only parts of it. By their very nature, local polynomials are more accurate within their local realm. It depends on our knowledge of what the data is supposed to represent, whether a single global

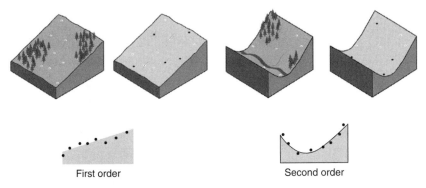

Figure 53 Polynomials of first and second order

polynomial is sufficient, or whether we need to subdivide the study area into regions (see Figure 54). Especially, lower degree polynomials are smooth interpolators – the resulting surface tends to miss measured values.

10.1.3 Splines

Splines are a common function in CAD packages, where the goal is to create smooth surfaces that minimize turbulence. The word originally described a long piece of wood that is bent by attaching weights along its length, pulling it into a desired shape (e.g. the outline of a violin or a ship's hull – see Figure 55).

Starting in the 1940s, mathematicians used the idea of weights pulling orthogonally to a line to develop rather complicated sets of local polynomials. They refer to splines as radial basis functions. The calculation of splines is computing intensive; the results definitely look pretty but may not be a good characterization of a natural landscape. Similar to IDW, the input points remain intact (see Figure 56) – which means that splines, in spite of their smooth appearance, actually are exact interpolators (see Figure 57).

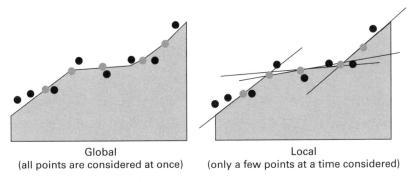

Global
(all points are considered at once)

Local
(only a few points at a time considered)

Figure 54 Local and global polynomials

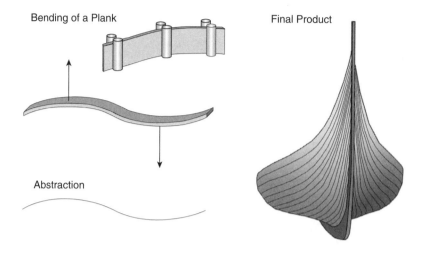

Figure 55   Historical use of splines

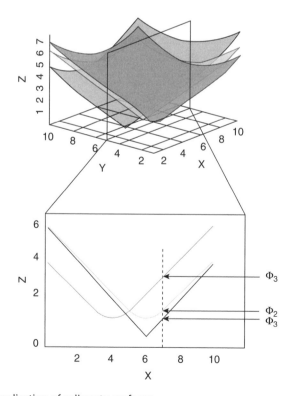

Figure 56   Application of splines to surfaces

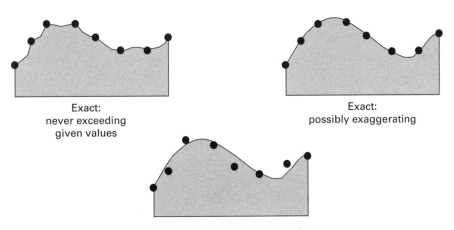

Exact:
never exceeding
given values

Exact:
possibly exaggerating

Figure 57   Exact and inexact interpolators

10.1.4 Kriging

All of the above interpolation methods use a model of what the analyst believes is the best representation of the interpolated surface. **Kriging** does this too; however, it uses statistics to develop the model. Originally developed for applications in the mining industry, it is now widely used in epidemiology, the environmental sciences, and of course geophysics. The term 'kriging' now signifies a whole family of methods, which to explain would go way beyond the scope of this book. The following is therefore only a general description of what underlies all kriging methods.

Kriging adopts the simple polynomial view of the world but concentrates on the tail end of the equation – called error – that describes what is *not* captured by the equation proper. In

$$y = a + bx + cx^2 + e$$

for instance, we have a second-degree polynomial with an error term e, which basically is a vessel for all discrepancies between the model result and the observed outcome (remember, polynomials are smooth or inexact interpolators). Using the First Law of Geography, it is now fair to assume that all the errors are spatially **autocorrelated** – that is, they increase the further we get away from a measured point. Kriging uses a brute force method of computing the relationships between *all* measured points and then runs statistics over the resulting table to determine which points have how much influence on what other points. This information is then fed back into the surface equation, which ideally is then error-free, making kriging an exact interpolator.

In practice there are a number of complications, each of which is addressed by a particular kind of kriging method. Especially the more sophisticated forms of kriging are extremely computing intensive; the results are no great shake if the number of original measurements is too small and the calculations run out of bounds if we have a rich input dataset. For the right number of points, and if the computing power is available, kriging delivers very robust results.

10.2 Spatial analysis

Spatial analysis comprises of a whole bag of different methods dealing with the quantitative analysis of spatial data. It ranges from simple geometric descriptors to highly sophisticated spatial interaction models. In a narrow sense, spatial analysis is the decidedly pre-GIS set of methods developed by geographers during the quantitative revolution of the 1960s, who in turn borrowed from analytical cartographers. Their goal was to describe geographic distributions, identify spatial patterns, and analyze geographic relationships and processes – all without GIS and often enough even without anything close to a computer.

10.2.1 Geometric descriptors

This is the application of descriptive statistics to spatial data. While we may sometimes borrow directly from traditional statistics, more often than not spatial also means special. In other words, we will have to come up with procedures that capture the spirit of what the traditional methods try to accomplish but adjust their implementation to the multi-dimensionality of spatial data, and possibly more important to the fact that we cannot assume spatial samples to be independent of each other.

We referred to the phenomenon of spatial autocorrelation above but did not expand on what a nuisance it poses for the statistical analysis of spatial data. Traditional statistics is based on the fact that samples are taken independently of each other and most methods assume a normal distribution. Both of these assumptions do not hold with geographic data – if they did, then there would be no basis for the discipline of geography. If a distribution is random then it is decidedly non-geographic, and Tobler's First Law of Geography would not hold.

The most basic descriptors in traditional statistics are mean, mode and standard deviation. Of these, the mean is relatively easy to translate into a spatial context; we just have to make sure that we calculate the average along as many dimensions (typically 1 to 3 for transects, areas or volumes) as we need. Figure 58 gives an example of a geometric mean.

The geometric median is a bit different from its traditional counterpart. Calculating the median values along x, y and possibly z to then mark the median location does not capture what we usually strive for in the calculation of a median. In traditional statistics, the median marks the point that is as far away from one end of the distribution as from the other. Translated into the spatial realm, this means that we are looking for a location that does the same not just within a dimension but also across. As it turns out, this is a really useful measure because it describes the location that minimizes the combined distances from that central point to all other locations. Unfortunately, there is no simple equation for that – the point can be found only through iterative optimization. Figure 59 illustrates the difference between a spatial mean and a spatial median.

A simple measure of central tendency is often too crude to adequately describe a geographic distribution. Analogue to the standard deviation in traditional statistics,

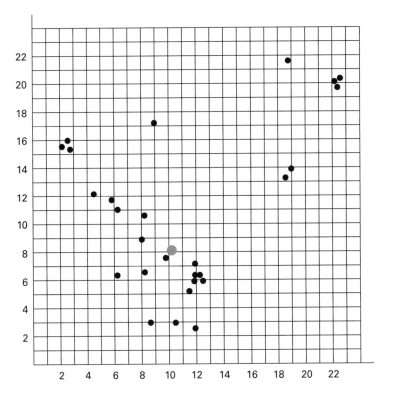

Figure 58 Geometric mean

we can employ standard distance as the circle around the mean, which captures a significant number of neighbors. The smaller the circle the more compact is the phenomenon; a wide circle tells us that the spatial mean is not very representative. However, by just calculating the standard distance, we throw away a lot of additional information. If we separate the standard distance into its *x* and *y* components, we get a standard deviational eclipse (see Figure 60) that tells us about the orientation or direction of the phenomenon and hence gives us additional clues as to what causes or at least influences it. This even applies to linear features, as a multitude of paths distributed over a larger surface (e.g. hurricanes or a river network) provides valuable clues as to what forces a change in direction.

The field of spatial pattern descriptors was expanded by landscape ecologists in the 1980s, who developed a myriad of measures to describe shapes and geometric relationships between features. **Shape measures** (see Figure 61) try to come up with characteristic numbers such as the ratio of edge length to area size, the degree of roundness (with a circle being perfectly round), or a figure for a feature's fractal dimension. While we will discuss more advanced spatial relationships in the next section, it is worthwhile to mention that a number of landscape ecological measures calculate average, min and max distance between features in general and on a feature class by feature class basis.

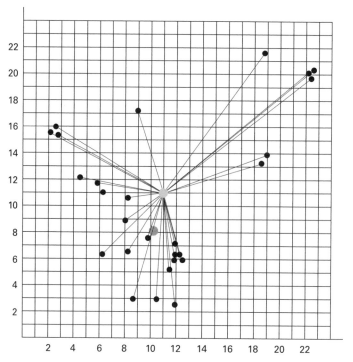

Figure 59 Geometric mean and geometric median

10.2.2 Spatial patterns

All of the above are global descriptors; they work well if the phenomenon we are studying is evenly distributed across the study area. But those are not the interesting geographic research questions. If we want to pursue locational variation then we need descriptors of local change. Although it sounds like an oxymoron, we have both global and local descriptors or local change.

Similar to the way we measure confidence in traditional statistics, we use the difference between observed distributions and expected (using the null hypothesis of randomness) to determine the degree of 'geography-drivenness'. Unfortunately, this is a bit more difficult than in traditional statistics because the numbers (and hence our confidence) change depending on how we configure size and origin of the search window within which we compare the two distributions.

One of the most often used spatial analytical methods is a nearest-neighbor analysis. Here we measure for each feature (zone in the raster world) the distance to its nearest neighbor and then calculate the average, min or max distance between neighbors of the same class or neighbors of two classes that we want to juxtapose with each other. Again, we can use a comparison between observed versus expected nearest-neighbor distance to in this case describe a particular distribution as clustered, random or dispersed. When we do this repeatedly with ever-changing search

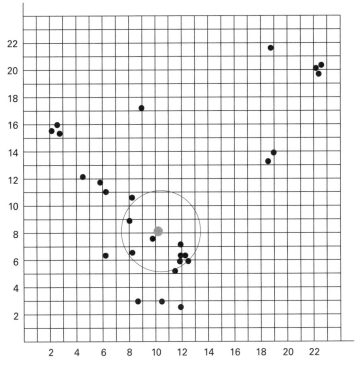

Figure 60 Standard deviational ellipse

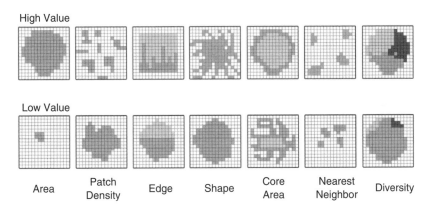

Figure 61 Shape measures

window size, we find the scale at which a given spatial pattern is particularly prominent, which in turn helps us to identify the process that is driving the spatial pattern.

So far, we have assumed that all features either belonged to one and the same class
or to a limited number of classes, for which we then describe the respective spatial

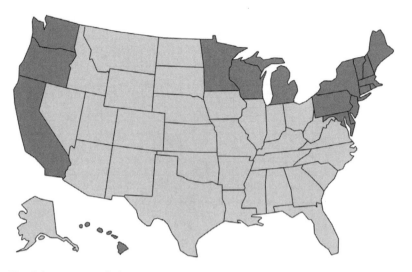

Figure 62 Joint count statistic

relationship. Alternatively, we could look at the attribute values of neighboring features and try to determine how similar or dissimilar they are. Common measures for the similarity of neighboring features are Geary's contiguity ratio c and Moran's I, and more recently the general G-statistic, which measures the degree of concentration of high versus low values. For categorical values, finally, we can use a so-called joint-count statistic, which compares the observed combination of neighboring values with a random combination. Figure 62 is a joint-count statistic of blue versus red states in the 2004 presidential elections in the United States.

All of the above is commonly applied to Euclidean distances but all of these measures work just as well on cost distances. And last but not least, the pattern analyzers can be applied to regional subsets, which often is more telling than a global measure.

10.2.3 The modifiable area unit problem (MAUP)

As mentioned at the beginning of the previous section, there are both global and local pattern detectors. The problem with the local ones, although they would be much more specific, is that it is hard to tell how to draw the boundaries. And as if this is not enough, more often than not the local boundaries for our data are predetermined. When we want to work with census data, for instance, we do not have control over how the data is collected or aggregated, and numerous studies have proven that, by drawing different boundaries, the results of a spatial analysis could be completely reversed. Without access to non-aggregated data, this is a severe limitation of spatial analysis, similar though not the same as the ecological fallacy problem in traditional statistics.

10.2.4 Geographic relationships

Another major contributor to spatial analysis techniques is the discipline of regional science, somewhat of a hybrid between economic geography and spatial econometrics. Some of the network-based location–allocation models come out of that realm, but what interests us here is the use of systems of regression equations to represent these relationships between geographic features. The polynomials that we encountered in geostatistics can be used the other way around – not to calculate missing values but to determine the underlying forcing functions that result in the observed values. Although there are examples for global regression analysis, the local (also known as geographically weighted) regression is of particular interest.

Many of these calculations are computationally very expensive, especially because an unbiased analysis requires the repeated run of many scenarios, where parameters are altered one at a time (Monte Carlo analysis). The frustration with this Pandora box of spatial analysis problems led to the development of geo-computation as a field, where the latest information science methods are applied to solving uncomfortably large spatial analysis problems. We look at these in the next chapter.

11 Geocomputation

Geocomputation is a set of computational methods that has been customized to address the special characteristics of spatial data. Given that definition, GIS would be a geocomputational method but it is decidedly not. The term was invented by geographer Stan Openshaw and became institutionalized with the first Geo-Computation conference in 1996. The term 'computational' has come to replace what used to be known as artificial intelligence techniques: from genetic algorithms, neural networks, and fuzzy reasoning to cellular automata, agent-based modeling, and highly parallelized processing. The common ground behind all of these is that if we throw lots of processing power and the latest advances in information science at large spatial datasets, then we have a good shot at deriving new insights.

A look at the proceedings of the GeoComputation conference series (www.geocomputation.org) conveys the wide range of topics, far more than could be covered in this chapter. We will concentrate here on five areas of research that have matured more than others: fuzzy reasoning, neural networks, genetic algorithms, cellular automata and agent-based modeling systems.

11.1 Fuzzy reasoning

As mentioned above, geocomputational techniques are borrowed from information science and then applied to spatial data. Lofti Zadeh (1965) formalized a set of rules to work with multi-valued logic, which allows us to capture the multi-valuedness of our thinking. Rather than categorizing everything as yes/no, black/white, zero/one etc., as we did when we introduced Boolean logic in Chapter 6, fuzzy logic extends the hard values zero and one to everything in between. An attribute can now be a little bit of green and a little more of blue rather than either green or blue. And applied to geographic phenomena, Florida can be 10% tropical, 80% subtropical, and another 10% moderate in climate.

The best everyday illustration of how fuzziness works is a shower knob that this author found in the bathroom of a New Zealand colleague (see Figure 63). People tend to have different opinions about what is warm water. By having the marking for cold water become ever thinner as the marking for warm water increases in width, the definition covers a wide range of beliefs with the majority somewhere in the middle at around 45°centigrade.

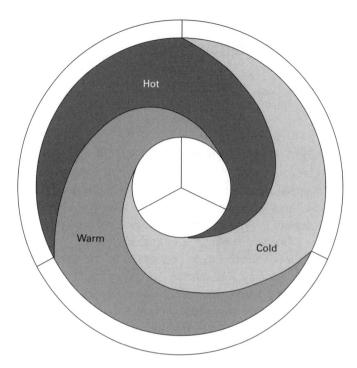

Figure 63 Shower tab illustrating fuzzy notions of water temperature

Formally, we describe fuzziness as a set of values ranging from 0 to 1. The grade to which an individual observation z is a member of the set is determined by a membership function, where membership values close to 1 represent observations close to the core concept. The value of the membership function reflects a kind of degree that is not based on probability but on admitted possibility. This concept of fuzziness allows us to work with imprecision, describing classes that for various reasons do not have sharply defined boundaries (Burrough and Frank 1996). The use of fuzzy sets is appropriate, whenever one has to deal with ambiguous, vague and ambivalent issues in models of empirical phenomena, and even supports working with qualitative data.

Data in fuzzy sets can be manipulated using basic operations that are similar to those found in Boolean logic – union (OR), intersection (AND) and negation (NOT). These operations are employed on both the spatial and the attributive aspects of an observation. The union operation (OR) combines fuzzy sets by selecting the maximum value of the membership function. The intersection operation (AND) requires the selection of the minimum membership value of the fuzzy sets in question. These operations perform the computation of a new membership value, which is called the joint membership function value.

The beauty of fuzzy logic applications in GIS is that it (a) overcomes the simplistic black/white perspective that traditional GIS forces us to adopt, and (b) it – at least in theory – allows us to work with qualitative notions of space. The Conference on Spatial Information Theory (COSIT) series is to a large degree devoted to the development of methods of qualitative spatial reasoning; unfortunately not much of the work presented there (1993–2005) has made it into readily available software.

11.2 Neural networks

With the advent of large spatial databases, sometimes consisting of terabytes of data, traditional methods of statistics such as those described in the previous chapter become untenable. The first group of GIScientists to encounter that problem was remote sensing specialists, and so it is no surprise that they were the first to 'discover' neural networks as a possible solution. Neural networks grew out of research in artificial intelligence, where one line of research attempts to reproduce intelligence by building systems with an architecture that is similar to the human brain (Hebb 1949). Using a very large number of extremely simple processing units (each performing a weighted sum of its inputs, and then firing a binary signal if the total input exceeds a certain level) the brain manages to perform extremely complex tasks (see Figure 64).

Feature vector

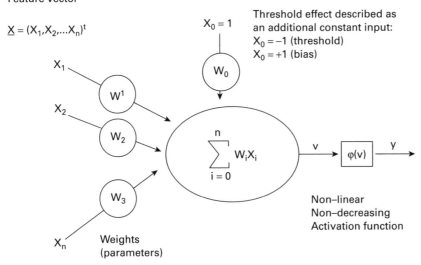

Figure 64 Schematics of a single neuron, the building block of an artificial neural network

Using the software (sometimes, though rarely, hardware) equivalent of the kind of neural network that makes up the brain, artificial neural networks accomplish tasks that were previously thought impossible for a computer. Examples include adaptive learning, self-organization, error tolerance, real-time operation and parallel processing. As data is given to a neural network, it (re-)organizes its structure to reflect the properties of the given data. In most neural network models, the term 'self-organization' refers to the determination of the connection strengths between data objects, the so-called neurons. Several distinct neural network models can be distinguished both from their internal architecture and from the learning algorithms that they use, but it would be beyond the scope of this book to go into detail here.

An important aspect of neural networks is whether or not they need guidance in learning. Based on the way they learn, all artificial neural networks can be divided into two learning types – supervised and unsupervised (analogous to the same idea in image classification used by remote sensers). In supervised learning, a desired output result for each input vector is required when the network is trained. It uses the target result to guide the formation of the neural parameters. In unsupervised learning, the training of the network is entirely data-driven and no target results for the input data vectors are provided. A neural network of the unsupervised learning type, such as Kohonen's (1982) self-organizing map, can be used for clustering the input data.

This alludes to the fact that the outcome of the application of neural networks is nothing really new. All this wizardry results in pretty much the same regression equations that we encountered in the previous chapter. There are two main differences. First, given the data volume, we could not have arrived at these results, which is the positive aspect. On the downside, the results are data and do not give us any insight into what is actually happening. From a scientific perspective, statistics is supposed to help us understand how things work. Neural networks, however, act like a black box – there is no algorithm (and no explanatory structure) that would help us to understand the phenomenon we are studying.

11.3 Genetic algorithms

There is nothing inherently spatial in genetic or evolutionary programming, so the reader might wonder why they became a popular geocomputational tool. Invented by Holland (1975), they are the dynamic equivalent of neural networks. While the latter are used when we have a large amount of data, genetic algorithms are used when we have a large number of possible solutions. A nice spatial example is the traveling salesman problem, where the task is to find the optimal sequence of customers in a sequential path. The problem cannot be solved for more than a handful of points because of the combinatorial explosion of options. This is, by the way, the reason why computers have not yet been able to beat a good player of the Japanese game of 'Go', another inherently spatial application. Genetic algorithms cannot claim to find the absolute best solution, but they are very good at finding better solutions than anyone or anything else.

I alluded to the use of genetic algorithms at the end of Chapter 7 (Location–Allocation), when we found that the model becomes intractably complicated. When we have a large number of origins and destinations, with multiple cases of each other influencing weights, then the equations not only become long and complicated, but the possible solution space of varying parameters becomes as large as in our traveling salesman problem, depicted in Figure 65. Who would venture a guess, which part of the equation should be tweaked to improve the result?

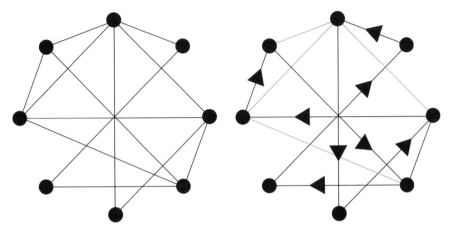

Figure 65   Genetic algorithms are mainly applied when the model becomes too complicated to be solved deterministically

Evolutionary programming starts by generating a population of purely random expressions – that is, random model equations. These are evaluated in terms of a fitness function. The best expressions are reused and sent to compete with a new generation of crossovers and slightly mutated versions of previously successful expressions. This process is repeated until no improvement is achieved. The terms 'crossover' and 'mutation' are borrowed from their biological analogues and function exactly the same way (see Figure 66). A crossover is a mixing of previously successful strategies, while a mutation is a slight alteration. Together with the best members of a previous generation, these new entities have to prove themselves. If they succeed (i.e., fare better in the evaluation of fitness for a particular goal), then they are allowed to stay for the next round.

Evolutionary techniques have not yet made it into commercially available GIS packages, but public domain versions of linkages are available. The interested reader may want to search www.sourceforge.net for a combination of the terms genetic and GIS.

11.4 Cellular automata

CA are a modeling framework for spatially continuous phenomena (Langton 1986), such as landscape processes or urban sprawl (Haff 2001; Box 2002; Silva and Clarke 2002). They are simple models used to represent the diffusion of things such as matter, information or energy, over a spatial structure. In its most simple form, a CA is

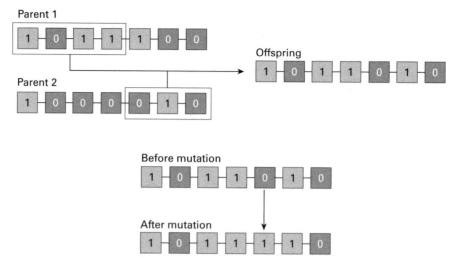

Figure 66 Principles of genetic algorithms

composed of a uniformly tessellated surface (typically a grid) whose cells may exist in a finite number of discrete states (see O'Sullivan 2001 for extensions). As such, CA can be considered a dynamic extension to raster GIS (Bian 2000). Each cell has an identically sized neighborhood consisting of nearby cells and a rule set defining how each cell changes based on the state of its neighborhood. These changes can be a function of either relative or absolute models of time, absolute time being where the scheduled tick of the model clock defines the change, and relative time expressed as a cascading process of event-based changes from one cell to the next. With these component parts, the model is initiated and run where each cell in the CA checks its neighborhood and changes its state based on the rules defining its behavior.

Despite the simplicity of construction, the dynamics of a CA model can produce complex results. For example, O'Sullivan measures change as the record of the time-series evolution of a measure of spatial pattern (2001).

However, CA are limited when it comes to modeling dynamic spatial phenomena. The most important limitation is that the structure of the tessellation is typically static, although there has been some promising experimentation with mutable CA in urban modeling (Semboloni 2000), and the use of self-modifying rules to capture nonlinear behavior (Silva and Clarke 2002). Yet there remains little scope for feedback and consequent self-organization of the cellular structure.

11.5 Agent-based modeling systems

Agent-based modeling (**ABM**), synonymous with individual-based modeling in ecology (Bian 2000), is a simulation methodology focused on mobile individuals and their interaction. It is based on the development of multi-agent systems (MAS), which were created in the field of distributed artificial intelligence (Gilbert and Terna 2000).

'Agent' is a generic term used for any constituent entity whose behavior we wish to model, and for its representation within the model. Agent-based models offer the ability to capture the dynamic interactions of individuals and the context in which they occur. Agent-based models enable the creation of 'artificial societies' which can be viewed as laboratories in which to conduct experiments (Epstein and Axtell 1996). Agents are defined, placed in an environment and given a set of bounded rules of behavior. The goal is to observe how interactions among individuals produce the collective behaviors that are being studied.

An agent-based simulation implemented in the framework of a computational laboratory offers the following advantages (Epstein and Axtell 1996; Gilbert and Terna 2000). First, agent-based models allow heterogeneity among individuals that more closely approximates the variety found in life. Second, the agents and the landscape can be held constant or systematically varied in order to provide a level of control impossible to attain using traditional social science methods. Third, the combination of heterogeneous agents and control enables the researcher to conduct a variety of experiments, using different conditions or applying various prevention scenarios and then evaluating outcomes for minimal cost compared to experiments in the real world. Using simulations allows us to repeat experiments under controlled conditions and to compress spatio-temporal scales, so we are no longer limited to observing just a few outcomes that happen to be presented by the real world. In addition, we are able to explore, evaluate and refine alternative scenarios and plans for remediation safely and at reasonable costs and risks.

The observables or attributes of an agent (including spatial location) are measurable characteristics of the agent that change over time (Parunak *et al.* 1998). These observables describe the state of the system at any one time and are the primary output of an ABM. ABMs develop histories of system states, where, as with temporal extensions to GIS, change is handled by storing the system state at each time or by storing vectors of events for each agent; that is, an agent logs each new state it enters. The focus of ABM is to understand the emergent outcome of each model, where emergent 'denotes the stable macroscopic patterns arising from the local interactions of agents' (Epstein and Axtell 1996, p. 35). In terms of spatial ABMs, this is the spatial pattern of observables (e.g. Parker and Meretsky 2004).

The primary distinction between CA and ABM is the conceptual primitive used to represent phenomena. In CA, this primitive is a static cell or pixel, a collection of which composes a layer of cells. Its dynamics involve each cell transferring information to its neighboring cells. An ABM, in contrast, is composed of distinguishable objects, the same geometric primitives of point, line or polygon data models found in GIS. Agent-based modeling enables the dynamic, situation-based decisions of individuals to drive emerging macro-level patterns of the phenomenon under study and is indispensable to the modeling of individual-level decision-making. Furthermore, an agent has the added advantage of being mobile.

There has been interest in intelligent software agents in GIScience in a variety of contexts. In particular, agents have been employed in geographic simulation modeling of a variety of phenomena, including simulation of land-use/cover change (Manson 2002; Parker *et al.* 2003), wayfinding (Raubal 2001) and social simulations (Gilbert and Doran 1994; Epstein and Axtell 1996; Gilbert and Troitzsch 1999; Conte *et al.* 2001). Only limited integration of GIS and agent-based models to simulate social phenomena has been achieved (Gimblett 2002). Agent-based simulation models are particularly promising for urban applications (Batty 2001; Benenson *et al.* 2002; Deadman and Gimblett 1994; Dean *et al.* 2000; Westervelt and Hopkins 1999).

12   Epilogue: Four-Dimensional Modeling

Many of the geocomputational techniques discussed in the Chapter 11 go beyond the scope of commercial GIS. That is partly because the tools are too complicated for a mass market, and partly because the problems these tools are applied to are too academic. Before the reader starts to abandon the concepts and techniques towards the end of this book as belonging into the ivory tower, I hasten to outline, with this final chapter, why the research frontier in GIScience is important to the general public. This chapter could be read as 'all the things you cannot (yet) do with GIS'.

Based on the last few chapters, the limitations of GIS should now be obvious. In most general terms, they can be described as the lack of currently easily available software to deal (a) with true 3-D, (b) with spatial processes, and (c) with qualitative data. Interestingly, (a) does not seem to pose a significant problem in the real world. As discussed in Chapter 9, true 3-D GIS have been developed for mineral resources applications (and arguably but not easily proven for military applications).

There are two distinct directions into which 3-D applications are moving. Based on market demand and increased graphics capabilities of modern hardware, 3-D visualization is becoming commonplace. It comes in the form of spherical representations of the globe, oblique scene rendering, fly-throughs, and even photographic textures of extruded buildings. For analytical purposes the development of true 3-D data structures is more interesting. There is an official standard in the form of the geographic markup language (GML v3), a pseudo standard in the form of U3D (ECMA 2006), which now that it is supported in the latest versions of general-purpose document viewing software (Acrobat 8) achieves wide market penetration, and terrain models that mix and match all three forms of 3-D data: TINs, DEMs and extruded vector data. We hence now finally have the data structures to support real 3-D analysis, and it is only a matter of time until the analysis methods will follow suit. One early example has been presented by Mennis *et al.* (2005) in the form of a multidimensional map algebra. The development of semantics for 3-D data that helps us to distinguish buildings from each other by more than just their geometry is part of the CityGML initiative (CityGML 2005).

The incorporation of qualitative data, and even more problematic, qualitative spatial relationships, has not progressed much in the past ten years. The emphasis in that research area has shifted towards the development of spatial ontologies in support of the semantic web, an intelligent classification of web-based data resources and already alluded to with respect to CityGML.

This leaves us the most promising area for new developments in GIScience – the realm of spatial process modeling. Cellular automata (CA) and agent-based models

(ABM) (Epstein and Axtell 1996; Gilbert and Terna 1999) were introduced in Chapter 11. They become even more interesting when they are run on richly structured landscapes. These bottom-up models focus on studying the emergent properties of systems by starting with individual-level interactions. They both model the same notion of underlying absolute space and utilize the same types of time, falling into what Zeigler terms discrete time or discrete event systems, depending on the modeling approach taken (Zeigler *et al.* 2000). The development of object-oriented software design has enabled scientists to develop models that realistically reflect the objects and relationships found in the real world (Gilbert and Terna 1999). And from a practical perspective, it allows software engineers to link GIS objects with ABM objects, as for instance in the AgentAnalyst extension that has been developed as a public domain project (http://www.institute.redlands.edu/agentanalyst).

On a more traditional side, especially environmental scientists (hydrologists, animal ecologists etc.) have for quite a while and with some success tried to nudge GIS to deal with time. As long as this is done within GIS, most attempts are based on map algebra. Some looping and conditional constructs, well-known from procedural programming languages, allow for state-based changes of features. The change in a landscape is then the sum of the changes of the features that it consists of (Pullar 2003). This, of course, does not capture transitions from one type of feature to another, such as when a cliff erodes to become a beach. One step further goes the PCRaster system developed at the University of Utrecht (The Netherlands), which addresses the needs of geophysicists and hydrologists to include differential equations in their GIS work.

Be it for lack of a commercial vendor or because a wider applicability has not been shown, many people interested in truly dynamic phenomena such as groundwater modeling (GMS, MODFLOW), wildfire spread (FARSITE), traffic congestions (EMME/2, WATSim) or weather forecasting (CALPUFF) prefer to link GIS with external software packages capable of dynamic modeling. What all of these packages lack (and why they link to GIS) is the notion of spatial differentiation. Space, and hence geography, is treated as a dependent variable if it is acknowledged at all.

In addition to traditional forms of GIS process modeling or the linking of GIS with external dynamic modeling programs, there is a third, and so far not much explored option: truly spatio-temporal systems that have processes as their building blocks. Many of the processes that we study in geography and related disciplines are the confluence of smaller scale (both spatial and temporal) processes. For instance, a housing boom can be the result of increased immigration, a disastrous hurricane, or the fact that other forms of capital investment are less lucrative. All of these are not features in the traditional GIS sense but processes.

A logical question then is: What can we expect to see from this form of process modeling in the near future? We will probably have a good number of process models, all well-specified, albeit in the beginning using different formalizations. The research agenda therefore includes development of a uniform process description language, similar to what the unified modeling language (OMG 2005) does for structures (UML 2 allows for the representation of activities but falls short of the needs of dynamic process modeling). Ideally, such a language would have the expressiveness and ease of use of the web ontology language OWL (McGuinness and van Harmelen 2004), while extending it to include rules and behavior. The Kepler system for scientific workflows is an early and still fairly primitive example for the kind of process libraries. The value of such process libraries has been recognized, both in the business world, where process models are a well-established component in operations research, and in the natural sciences (see, for example, the Kepler system (http://keplerproject.org) that is part of the SEEK program heavily funded by NSF).

Linking these kinds of process model with 3-dimensional GIS models will be the ultimate goal. Unfortunately, this book will be long out of print before we can expect such software systems in the hand of the reader.

Glossary

ABM
Agent-based modeling system – a simulation tool used to investigate the aggregate outcome of actions of individuals

Accuracy
The difference between what is supposed to be encoded and what actually is encoded

Address
Short for street address, a spatial reference commonly used by postal services and humans but not by GIS

Attribute
The characteristic of a feature or location

AutoCAD™
Popular computer design program whose drawing exchange format has become a de-facto standard for the exchange of geometric data

Autocorrelated
Statistical fact underlying most geographic phenomena that renders traditional statistical techniques obsolete

Boolean logic
Binary logic underlying most digital equipment; also commonly used in GIS overlay operations

Buffer
Result of a GIS operation determining the neighborhood of a feature

CA
Cellular automata implement transition rules to mimic the evolution of an artificial landscape

CAD
Computer-aided design – a type of software that processes geometries similar to GIS but at a larger scale, without geo-reference and less emphasis on the link between geometries and attributes. A number of GISystems have been developed from CAD software

Centroid
Middle-most (central) point of an area or region

Coordinate
Location in a Cartesian or polar coordinate system

dBASE™
Originally a database program, it has become a de-facto standard for the exchange of attribute data

DEM
Digital elevation model – a framework for recording spot elevations in a raster layer

Digital number
Attribute value of a cell in a raster image

Digitizing
The act of transforming analogue data (such as a paper map) into digital data

Feature
The object of interest in a GIS; it has to have a location and some attribute

Field view
Represents space as a continuous surface of attributes

First Law
An off remark in a 1970 article that became famous (... of Geography, because it indeed underlies everything geographical)

Focal function
Neighborhood function in map algebra

FTP
File transfer protocol – a network standard that is commonly used for the transfer of large datasets

Fuzzy reasoning
A form of multi-valued logic based on set theory that allows for reasoning with vague data and relationships

Geodemographics
A spatial analysis of demographic data pioneered in marketing applications

Geographic object
See Feature

GeoTIFF
Tiff format with a spatial reference

GIS
Geographic Information System

GIScience
Body of knowledge created by combining many of the mother disciplines that are necessary for the successful development of GIS

GML
Geographic markup language – an XML dialect for the exchange of data between GISystems

GPS
Global positioning system – an array of satellites that (given an appropriate receiver) can help to determine one's location on Earth

IDW
Inverse distance weighting – a spatial interpolation method that incorporates information from known points according to the inverse of their distance to the unknown point

ISO
International Standards Organization – instrumental in setting many of the standards (such as 19115) used for the processing of geographic information

Kriging
Spatial interpolation method that uses weights based on the statistical analysis of covariances in a global point dataset

Lineage
The history of a dataset – an important metadata item

Map algebra
Extremely powerful rule set for combining raster layers

Map projection
The application of mathematical formulas to transform spherical coordinates describing features on the surface of the Earth to Euclidean coordinates used in most GIS

MapQuest®
Company that pioneered the use of online mapping

MAUP
Modifiable area unit problem – arising from the attempt to combine sets of data that have been aggregated in different though overlapping spatial units

Metadata
Literally data about data – important for archiving and re-use of geographic information

Neural network
Computational technique that mimics brain functions to arrive at statistical results

Object view
Represents the features as discrete objects with well-define boundaries in an empty space

Ontology
Formal specification of the meaning of a datum

Overlay
Quintessential GIS operation that determines spatial coincidences

PDF
Portable document format – an open file format for the description of device-independent documents

Pedologist
One who studies soil science

Precision
The amount of detail that can be discerned in geographic information

Projection
See Map projection

Raster
Spatial organization of data similar to an array or a spreadsheet; space is completely filled by the cells that make up the raster

Regional science
Academic discipline at the intersection of economics and geography that developed its own set of spatial analysis techniques

Remote sensing
The technique (and science behind) gathering information from objects without touching them

Scanning
An automated form of digitizing that results in raster data

Semantics
The meaning of a datum

Shape measures
Set of statistical measures to describe spatial configurations; originally developed in landscape ecology

Spatial reference
Descriptor for a location on Earth

SQL
Structured query language – a standard (with many variations) way of querying a database

SVG
Scalable vector graphics – an XML dialect for the description of vector data

Thiessen polygon
For a point dataset, the area around one point that is closer to this point than to any other point

TIFF
Tagged image file format – an error-free storage format for raster data

TIN
Triangulated irregular network – a representation of a surface derived from irregularly spaced sample points

Topology
Branch of mathematics that deals with qualitative spatial relations. Topological relationships are important for many GIS operations and have been used as a check for the geometric consistency of a GIS database

UML
Unified modeling language – an ISO standard for the specification of database schemas

Unix
Family of multi-user operating systems

UTM
Universal Transverse Mercator projection and coordinate system. Originally used by the US armed forces, it is now common throughout the world for GIS applications covering larger areas

Vector GIS
GIS that uses points, lines and polygons to represent geographic features

Web 2.0
A set of techniques associated with web technologies that enable users to develop their own applications

XML
Extensible markup language – a superset of what many know as web description languages such as HTML. XML is not meant to be read by humans but to facilitate automated exchanges between computers

References

Alonso, W. (1978). A theory of movements. In Hansen, N.M. (ed.): *Human Settlement Systems: International Perspectives on Structure, Change and Public Policy*, pp. 197–211. Cambridge, MA: Ballinger Publishing.

Batty, M. (2001). Cellular dynamics: modelling urban growth as a spatial epidemic. In Fischer, M. and Leung, Y. (eds): *GeoComputational Modelling: Techniques and Application*, pp. 109–141. Berlin: Springer-Verlag.

Benenson, I., Omer, I. and Hatna, E. (2002). Entity-based modeling of urban residential dynamics: the case of Yaffo, Tel Aviv. *Environment and Planning B: Planning and Design*, **29**, 491–512.

Bettinger, P. and Sessions, J. (2003). Spatial forest planning: to adopt, or not to adopt? *Journal of Forestry*, **101**(2), 24.

Bian, L. (2000). Object-oriented representation for modelling objects in an aquatic environment. *International Journal of Geographic Information Science*, **14**, 603–623.

Box, P. (2002). Spatial units as agents: making the landscape an equal player in agent-based simulations. In Gimblett, H.R. (ed.): *Integrating Geographic Information Systems and Agent-based Modeling Techniques for Simulating Social and Ecological Processes*, pp. 59–82. New York: Oxford University Press.

Bryant, R. (2005). Assisted GPS: using cellular telephone networks for GPS anywhere. *GPS World*, May 2005, 40–46. Online article at www.gpsworld.com/gpsworld/content/print ContentPopup.jsp?id=163593

Burrough, P.A. (1986). *Principles of Geographic Information Systems*. Oxford: Oxford University Press.

Burrough, P.A. and Frank, A.U. (1996). *Geographic Objects with Indeterminate Boundaries*. London: Taylor & Francis.

Chrisman, N. (2002). *Exploring Geographic Information Systems*. New York: Wiley.

CityGML (2005). Exchange and storage of virtual 3D city models. OGC discussion paper, available online at https://portal.opengeospatial.org/ files/?artifact_id= 16675

Conte, R., Edmonds, B., Moss, S. and Sawyer, K. (2001). Sociology and social theory in agent-based social simulation: a symposium. *Computational and Mathematical Organization Theory*, **7**(3), 183–205.

Couclelis, H. (1982). Philosophy in the construction of geographic reality: some thoughts upon the advent of the 'qualitative revolution' in human geography. In Gould, P. and Olsson, G. (eds): *A Search for Common Ground*, pp. 105–138. London: Pion.

Couclelis, H. (1992). People manipulate objects (but cultivate fields): beyond the raster–vector debate in GIS. In *Theories and Methods of Spatio-Temporal Reasoning in Geographic Space*, vol. 639. Berlin: York: Springer-Verlag.

Deadman, P. and Gimblett, H.R. (1994). The role of goal-oriented autonomous agents in modeling people–environment interactions in forest recreation. *Mathematical and Computer Modeling*, **20**(8), 121–133.

Dean, J.S., Gumerman, G.J., Epstein, J.M., Axtell, R.L., Swedlund, A.C., Parker, M.T. and McCarroll, S. (2000). Understanding Anasazi cultural change through agent-based modeling. In Kohler, T.A. and Gumerman, G.J. (eds): *Dynamics in Human and Primate Societies*, pp. 179–205. Oxford: Oxford University Press.

ECMA (2006). *Universal 3D File Format*, 3rd edn. ECMA International and European Association for Standardizing Information and Communication Systems, Geneva. Online document accessible at www.ecma-international.org/ publications/files/ECMA-ST/ECMA-363.pdf

Egenhofer, M. (1993). A model for detailed binary topological relationships. *Geomatica*, **47**(3/4), 261–273.

Epstein, J.M. and Axtell, R. (1996). *Growing Artificial Societies: Social Science from the Bottom Up*. Cambridge, MA: MIT Press.

Gilbert, N. and Doran, J. (eds) (1994). *Simulating Societies: The Computer Simulation of Social Phenomena*. London: UCL Press.

Gilbert, N. and Troitzsch, K.G. (1999). *Simulation for the Social Scientist*. Buckingham: Open University Press.

Gilbert, N. and Terna, P. (2000). How to build and use agent-based models in social science. *Mind & Society*, **1**, 57–72.

Gimblett, H.R. (ed.) (2002). *Integrating Geographic Information Systems and Agent-based Modeling Techniques for Simulating Social and Ecological Processes* [Sante Fe Institute Studies in the Sciences of Complexity]. New York: Oxford University Press.

Gregory, I. (2003). *A Place in History: A Guide to Using GIS in Historical Research*. Oxford: Oxbow Books.

Haff, P.K. (2001). Waterbots. In Harmon, R.S. and Doe, W.W. (eds): *Landscape Erosion and Evolution Modeling*, pp. 239–275. New York: Kluwer.

Hamil, D. (2001). Your mission, should you choose to accept it: project management excellence. Online article at http://spatialnews.geocomm.com/features/mesa1/hamil1.pdf

Hebb, D.O. (1949). *The Organization of Behavior: A Neuropsychological Theory*. New York: Wiley.

Holland, J.H. (1975). *Adaptation in Natural and Artificial Systems: An Introductory Analysis with Applications to Biology, Control, and Artificial Intelligence*. Ann Arbor: University of Michigan Press.

Kirby, K. and Pazner, M. (1990). Graphic map algebra. In Brassel/ Kishimoto (eds): *Proceedings of the 4th International Symposium on Spatial Data Handling*, vol. 1, pp. 413–422. Columbus, Ohio, USA: International Geographical Union (IGU), Commission on Geographic Information Systems.

Kohonen, T. (1982). Self-organized formation of topologically correct feature maps. *Biological Cybernetics*, **43**, 59–69.

Langton, C. (1986). Studying artificial life with cellular automata. *Physica D*, **22**, pp. 120–149.

Maantay, J. and Ziegler, J. (2006). *GIS for the Urban Environment*. Redlands, CA: ESRI Press.

Manson, S.M. (2002). Integrated assessment and projection of land-use and land-cover change in the Southern Yucatán Peninsular Region of Mexico. In Parker, D.C., Berger, T. and Manson, S.M. (eds): *Agent-based Models of Land-Use and Land-Cover Change*, pp. 56–59. Bloomington, IN: LUCC International Project Office.

McGuinness, D.L. and van Harmelen, F. (2004). OWL web ontology language overview: W3C recommendation 10 February 2004. Available at www.w3.org/ TR/owl-features

Mennis, J., Leong, J. and Khanna, R. (2005). Multidimensional map algebra. In *Proceedings of the 8th International Conference on GeoComputation*, 1–3 August, Ann Arbor, MI. Ypsilante, Michigan, USA: Institute for Geospatial Research & Education.

Mitchell, M.L. and Jolley, J.M. (2001). *Research Design Explained*, 4th edn. Pacific Grove, CA: Wadsworth.

OMG (2005). *Introduction to PMG's Unified Modeling Language*. Needham, MA: Object Management Group. Online document, accessible at www.omg.org/gettingstarted/ what_is_uml.htm

O'Sullivan, D. (2001). Exploring spatial process dynamics using irregular cellular automaton models. *Geographical Analysis*, **33**, 1–18.

Parker, D., and Meretsky, V. (2004). *Measuring Pattern Outcomes in an Agent-Based Model of Edge-effect Externalities using Spatial Metrics*. Agriculture, Ecosystems, and Environment, **101**: 233–250.

Parker, D.C., Manson, S.M., Janssen, M.A., Hoffmann, M.J. and Deadman, P. (2003). Multi-agent systems for the simulation of land-use and land-cover change: a review. *Annals of the Association of American Geographers*, pp. 314–337.

Parunak, H.V.D., Sauter, J. and Clark, S.J. (1998). Toward the specification and design of industrial synthetic ecosystems. In Singh, M.P., Rao, A. and Wooldridge, M.J. (eds): *Intelligent Agents IV: Agent Theories, Architectures, and Languages* [Lecture Notes in Artificial Intelligence 1365], pp. 45–59. Berlin: Springer. Available at www.altarum.net/~van/agentDesignMethod.pdf

Pullar, D. (2003). Simulation modelling applied to runoff modelling using MapScript. *TGIS*, **7**(2), 267–283.

Raubal, M. (2001). Ontology and epistemology for agent-based wayfinding simulation. *International Journal of Geographical Information Science*, **15**(7), 653–665.

Reggiani, A. (2001). *Spatial Economic Science: Frontiers in Theory and Methodology*. Berlin: Springer-Verlag.

Semboloni, F. (2000). The growth of an urban cluster into a dynamic self-modifying spatial pattern. *Environment and Planning B*, **27**, 549–564.

Silva, E.A. and Clarke, K.C. (2002). Calibration of the SLEUTH urban growth model for Lisbon and Porto, Portugal. *Computers, Environment and Urban Systems*, **26**, 525–552.

Tobler, W.R. (1970). A computer movie simulating urban growth in the Detroit Region. *Economic Geography*, **46**, 234–240.

Tomlin, C.D. (1990). Geographic information systems and cartographic modeling. Englewood Cliffs, NJ: Prentice-Hall.

Wesseling, C. and van Deursen, W. (1995). A spatial modelling language for integrating dynamic environmental simulations in GIS. In *Proceedings of the Joint European Conference and Exhibition on Geographical Information*, (JEC-GI), vol. 1, pp. 368–373. Lisbon, Portugal: European Unbrella Organisation for Geographic Information.

Westervelt, J.D. and Hopkins, L.D. (1999). Modeling mobile individuals in dynamic landscapes. *International Journal of Geographical Information Science*, **13**(3), 191–208.

Wilson, J.P. and Gallant, J.C. (2000). *Terrain Analysis: Principles and Applications*. New York: Wiley.

Zadeh, L. (1965). *Fuzzy Logic and its Applications*. New York: Academic Press.

Zeigler, B.P., Praehofer, H. and Kim, T.G. (2000). *Theory of Modeling and Simulation: Integrating Discrete Event and Continuous Complex Dynamic Systems*. San Diego: Academic Press.

Index

raster
-based elevation models, 60–62
-based programs, 58
cells, **52**, 52–53, 59
data, 22, 51, 60, 65, 91
GIS, 32, 41, 49, 51, 61, 66, 81
zones, 53
rated space, *4*
real space, *4*
recoding, 29, **30–41**, 2–9
regional science, 45, 49, 75, 91
regression analysis, 75, 80
remote sensing, 1, 5–7,
62, 79, 91
resolution, 7, 18

sampling, 3–5
satellite imaging, 6–7, 13, 16, 64
scanning, 8, 91. See also
digitizing
scripts, 58
selection, 24–25, **25**
self-organization, 79–80, 82
semantics, 12, 91
sensitivity, 7
sensors, 6–7
shape measures, 71, **73**, 92
shortest-path analysis, 45–46. *See
also* optimization
slope, 62
space, **4**
spatial analysis, 21, 46, 65, 70,
72, 74–75, 92
spatial autocorrelation, 70
spatial Boolean logic, 40, **40**
spatial data, 1–2, 8, 17–19, 29
spatial distributions, 5, 9
spatial econometrics, 75
spatial interpolation, 65
spatial patterns, 72–74
spatial reference, 2, 21, 90–92
spatial relationships, 29,
32, **33**, 36
spatial search, 21, 39
space, types of, *4*
splines, 67, **67–68**
SQL (structure query language),
23, 92
standard deviational ellipse, **72**
standard space, *4*

statistics, traditional, 70, 72, 74
structure query language
(SQL), 24, 92

tagged image file format (TIFF),
91–92
Thiessen polygon, 44, **44**, 92
third dimension
digital elevation models and, 62
representation of, **59**
TIFF (tagged image file format),
91–92
TIN. *See* triangulated irregular
networks
Tobler's First Law of Geography, 46,
56, 65, 69–70
topology, 18, 34–36, **35**, 92
traveling salesman
problem, 80–81
triangulated irregular networks (TIN),
44, **60,** 60–61, 65, 85, 92
triangulation, 8

UML. *See* unified modeling
language
unified modeling language
(UML), 86–87, 92
Universal Transverse
Mercator, 15, 92
Unix, 16, 92
US
Census Bureau, 16
Geological Survey, 15–16
UTM (Universal Transverse Mercator),
15, 92

value grids, **58**
variable source problem, **5**
vector
-based GIS, 1, **3**, 45, 51,
58–60, 92
data, 6, 60, 86, 92
viewshed analysis, 61, **61**
visibility analysis, 61
Voronoi diagram. *See* Thiessen
polygon

Web 22, 92
web
-based geographic data, 16